Electric Power Conversion Handbook

Electric Power Conversion Handbook

Edited by Helena Walker

STATES
ACADEMIC PRESS
www.statesacademicpress.com

States Academic Press,
109 South 5th Street,
Brooklyn, NY 11249, USA

ISBN: 978-1-63989-171-9

Cataloging-in-Publication Data

Electric power conversion handbook / edited by Helena Walker.
    p. cm.
Includes bibliographical references and index.
ISBN 978-1-63989-171-9
1. Electric power production. 2. Direct energy conversion. 3. Energy conversion. 4. Electric inverters.
5. Electric transformers. I. Walker, Helena.
TK2896 .E44 2022
621.042--dc23

For information on all States Academic Press publications
visit our website at www.statesacademicpress.com

STATES
ACADEMIC PRESS

Contents

Permissions

List of Contributors

Index

Preface

The purpose of the book is to provide a glimpse into the dynamics and to present opinions and studies of some of the scientists engaged in the development of new ideas in the field from very different standpoints. This book will prove useful to students and researchers owing to its high content quality.

The rate per unit time at which the transfer of electrical energy takes place through an electric circuit is known as electric power. It is usually generated by electric generators but can also be supplied by some other sources such as electric batteries and solar panels. One form of electric energy can be converted into another form and the process is known as power conversion. The electrochemical or electrical device used for the conversion of electrical energy is referred to as power converter. It can convert direct current into alternating current and vice-versa. A power converter can also change the frequency or voltage of the current. The ever growing need of advanced technology is the reason that has fueled the research in the field of electric power conversion in recent times. Most of the topics introduced in this book cover new techniques and the applications of electric power conversion. Scientists and students actively engaged in this field will find it full of crucial and unexplored concepts.

At the end, I would like to appreciate all the efforts made by the authors in completing their chapters professionally. I express my deepest gratitude to all of them for contributing to this book by sharing their valuable works. A special thanks to my family and friends for their constant support in this journey.

Editor

Electrodeposition Approaches to Deposit the Single-Phase Solid Solution of Ag-Ni Alloy

Brij Mohan Mundotiya, Wahdat Ullah and Krishna Kumar

Abstract

The Ag-Ni films have attracted the attention of material scientists and researchers due to its applications as magnetic materials, catalyst, optical materials, etc. In the bulk, the formation of a single-phase solid solution of the Ag-Ni alloy film is difficult due to the large differences in the atomic size and the positive enthalpy of mixing. However, the immiscibility in the Ag and Ni constituents is diminished in the nano- sized level. The electrodeposition method has established itself as a most suitable method for synthesis of single-phase Ag-Ni alloy films due to its time efficiency, cost-effectiveness, and ability to mass production of single-phase solid solutions. In this method, the miscibility of alloying elements (i.e., Ag and Ni) and the quality of the Ag-Ni film can also be easily controlled by tuning the electrodeposition process parameters such as the magnitude of the applied current density, temperature of the electrolyte, additives in electrolytes, etc. This chapter presents a detailed overview of the process parameters affecting the miscibility, morphology, and the quality of the single solid solution of Ag-Ni film. Furthermore, the nucleation and growth mechanism of Ag-Ni film and the effect of the curvature of the deposited film's particles in the miscibility of the Ag and Ni elements have also been discussed in detailed.

Keywords: deposition parameters, electrodeposition, immiscibility, morphology, single-phase Ag-Ni alloy

Introduction

In recent years, the researchers are looking with the hope that the metastable (non-equilibrium) alloys might be materials with potentially unique and exploitable functional properties. The alloys in the non-equilibrium state are generally made from the immiscible element system such as Co-Ag, Ag-Ni, Au-Co, Ag-Pt, Fe-Cu, Co-Cu, and Ag-Fe alloy systems. In bulk state, constituent elements of these systems exhibit strong immiscibility behavior even up to very high temperatures. However, miscibility of their constituent elements can be enhanced by mixing them in nanoscale level to form a nano-size system. The enhanced miscibility of constituent elements in the nano-size system also induces its exploitable functional properties due to the evolution of the unique type of microstructure [1]. In recent years, numerous amorphous and nanocrystalline single solid solution alloy systems of immiscible elements Ni-Co [2], Ni-W [3], Ni-Si [4], and Ni-Fe [5, 6] have all been explored and reported in scientific literatures. Among the metastable alloys systems, Ag-Ni alloy system is one of the very interesting system due to the presence of one magnetic (Ag) and one nonmagnetic (Ni) alloying elements and their possible applications in the field of catalysts [7], magnetic materials [8, 9], electrical materials [10], and optical materials [11]. But due to the alloying characteristics of the constituent elements of the Ag-Ni system, these systems have been explored very little.

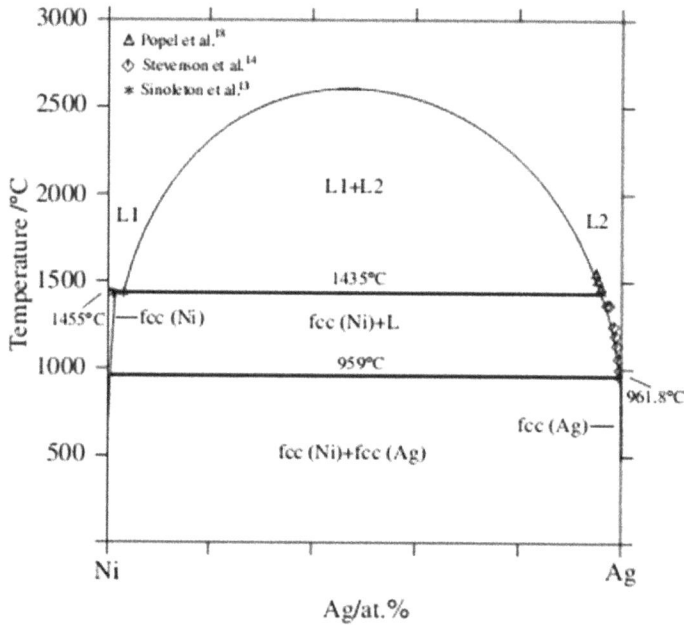

Figure 1. *Phase diagram of Ag-Ni system [15].*

The alloying elements of Ag-Ni system (i.e., Ag and Ni elements) have the same lattice structure (FCC), and the difference between their atomic sizes is very large (~14%). They are immiscible in both the solid and liquid states or even up to very high temperatures. The phase diagram of the Ag-Ni system is very simple (see **Figure 1**). As it is obvious from the phase diagram that Ag and Ni are mutually insoluble, even at 400°C, the solubility of Ni in Ag is only ~0.0219 at% [12]. The reasons of the high immiscibility of Ag and Ni alloying elements are (a) the large difference in their atomic sizes (~14%) and (b) a positive enthalpy of mixing (ΔH^{mix} for solid solution equi-atomic Ag-Ni alloy is 23 kJ/mol) [13, 14]. However, an enhancement of miscibility between Ag and Ni atoms in nano-size level to form a nano-size system has been achieved by researchers [8–11]. Enhancement of miscibility of atoms of nanoparticles of Ag and Ni can be attributed to (a) the effect of increase of particles curvatures with the reduction of their sizes that facilitate miscibility in accordance with the Gibbs-Thomson effect and (b) the decrease of the driving force for the nucleation and growth of the second phase within a nano-sized particle [13].

In recent years, the single-phase Ag-Ni alloys have successfully synthesized by various techniques such as ion beam mixing [16], laser ablation [10], gamma-ray irradiation [17], etc. Zhang et al. [17] synthesized the single solid solution phase of Ag-Ni alloy by simultaneous reduction of Ag and Ni ions by gamma-ray irradiation. The authors obtained a single solid solution phase of the Ag-Ni system with the homogeneous distribution of compositions of Ag and Ni atoms inside the particles. The average size of the particle of a single solid solution of Ag-Ni system was 6 nm. In another study reported by Kumar et al. [18], an average particle size of 30 nm of a single solid solution of Ag-Ni system was synthesized by immersion of a film of thermally evaporated fatty acids sequentially in solutions containing Ag^+ ions and Ni^+ ions. Srivastava et al. [19] studied the size-dependent micro- structural evolution for Ag-Ni alloy nanoparticles. The authors synthesized Ag-Ni nanoparticles by the co-reduction of Ag and Ni metal precursors. It was found that the Ag and Ni atoms were completely mixed in Ag-Ni alloy nanoparticles when the particle size was less than 7 nm. As the size of the particles was increased, the nanoparticles showed two separate regions: (a) a pure Ag region and (b) a solid solution of Ag-Ni alloy.

The electrodeposition method has established itself as a most suitable method for synthesis of single-phase Ag-Ni alloy films due to its time efficiency, cost-effectiveness, and ability to mass production of single-phase solid solutions [13, 20–26]. In the electrodeposition process, the apparatus used for the deposition of the film is less complicated than the vacuum-based methods such as molecular beam epitaxy or sputtering. In this method, miscibility of the alloying elements (Ag and Ni) to form a single-phase solution of Ag-Ni films and their quality can also be easily controlled by tuning the electrodeposition process parameters such as current density, electrolytes, electrolyte temperatures, additives in electrolytes, etc. In the following sections of this chapter, the overview of information related to the effects of these parameters on the miscibility of the alloying elements (Ag and Ni) and the quality of the deposited film are discussed.

Nucleation and growth mechanism of film

The deposition of the alloy films on the conductive substrate involves the nucleation and growth process. The nucleation process is very important because it decides the nature of the evolution of microstructure and crystallinity in the deposited alloy film. The alloy film deposition on the conductive substrate by electrodeposition method involves a heterogeneous process. The interface between the electrolyte solution and the substrate surface provides preferential sites for nucleation. As an example, consider the nucleation of a nucleus on the substrate surface (see **Figure 2**). On the substrate surface, nucleus forms in the shape of a half convex lens (spherical cap). The surface tension force acting on the surface- nucleus-electrolyte interfaces is determined by the formula given in Eq. (1). This equation is known as Young's equation [27]:

$$\gamma_{sl} = \gamma_{fs} + \gamma_{lf} cos\theta \tag{1}$$

where γ_{lf}, γ_{fs}, and γ_{sl} are the surface or interface energy of electrolyte-nucleus, nucleus-substrate, and substrate-electrolyte interfaces, respectively, and θ is the contact angle that depends on nature of the surfaces involved to form interfaces.

In the heterogeneous nucleation process, the probability of the nucleation is determined by the critical nucleus radius r* and the activation energy barrier $\Delta G*$ that a nucleation process must be overcome to form a nucleus. The formulae for the r* and $\Delta G*$ are given by Eqs. (2) and (3), respectively [28]:

$$r^* = \left(\frac{2\pi\gamma_{lf}}{\Delta G_l}\right)\left(\frac{sin^2\theta \cdot cos\theta + 2\,cos\theta - 2}{2 - 3\,cos\theta + cos^3\theta}\right) \tag{2}$$

$$\Delta G^* = \frac{16\pi\gamma_{lf}}{3\,(\Delta G_l)^2}\left(\frac{2 - 3\,cos\theta + cos^3\theta}{4}\right) \tag{3}$$

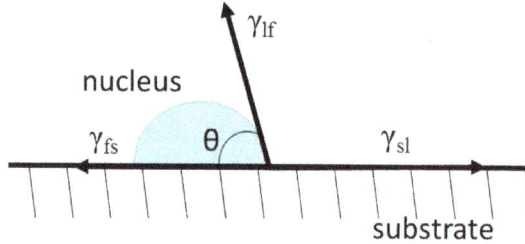

Figure 2. *Schematic diagram of a nucleus on a solid surface (substrate) with energy vectors and contact angle.*

where ΔG_l is the Gibbs free change per unit volume. If the radius (r) of the nucleus is smaller than the critical nucleus radius (r*), the nucleus is not survived for further growth and redissolved.

From the atomic point of view, the electrodeposition of film occurs in two different stages. In the first stage, few metal atoms from the electrolyte solution adhere to the substrate. Normally, it is considered that the first stage may process via either of the two mechanisms: (a) step-edge site ion transfer or (b) terrace site ion transfer [29]. In the step-edge site ion-transfer mechanism, the metal adions transfer from the electrolyte solution to a kink site of a step edge of the substrate surface via two paths. In the first path, the metal adions directly transfer from the electrolyte solution to the kink site of a step edge of the substrate. While in the second path, the metal adions diffuse from the electrolyte solution to the step edge first and then finally transfer to the kink site. On the other hand, in the terrace ion-transfer mechanism, the metal adions transfer from the electrolyte solution to the flat surface of the terrace region. The bonds between the metal adions are stronger to each other than to the surface of terrace regions. Therefore, metal adions easily diffuse from the terrace and transfer to the step-edge site and then finally to a kink site. These clusters of metal adions on the different sites of the substrate form nuclei of ordered structure.

Following the first stage, the second stage of growth of the nucleus occurs. The experimental observations revealed that there are three basic kinds of growth modes of alloy film [30]: (a) island growth, (b) layer growth, and (c) island-layer growth. When the bound between the depositing species (i.e., metal adions) is stronger than to the substrate, the island growth occurs. As the deposition proceeds, the islands of deposit subsequently grow and merge to form a continuous film on the substrate. The layer growth of film occurs, when the depositing species (i.e., metal adions) are bound strongly to the substrate than to each other. In the layer growth mode, first of all, a complete monolayer of depositing species is formed on the substrate, then a second layer starts to deposit on the monolayer. The islandlayer growth mode of alloy film is a combination of both island and layer growth modes. In the island-layer growth mode, first of all, a layer of deposit is grown on the substrate by the layer growth mechanism; this layer has some degree of residual stresses when the critical thickness of the layer is reached then due to an increase of the stresses on it, the island growth mechanism is activated.

Electrodeposition solution for Ag-Ni films

Table 1. *Electrodeposition parameters used for the deposition of Ag-Ni alloy films/nanoparticles.*

Coating/nano-particles	Electrolyte bath	pH	Temperature (°C)	Current density (mA/cm²)	Substrate	Reference
Ag-Ni film	$C_6H_5Na_3O_7$: 0.3 M $NiSO_4$: 0.7 M $AgNO_3$: 2–5 mM	4.5–7.5	Room temperature	0.5–50	Brass plate	Eom et al. [20]
Ni-Ag coating	$NiSO_4 \cdot 6H_2O$: 49.93 g/L $AgNO_3$: 0–0.339 g/L $C_6H_5Na_3O_7 \cdot 2H_2O$: 0–41.160 g/L H_3BO_3: 12 g/L	5.6	30	5	Cu plate	Raghupathy et al. [21]
Ni-Ag layer	$C_6H_5Na_3O_7$: 0.26 mol/L $NiSO_4$: 0.7 mol/L $AgNO_3$: 0.002 mol/L	5.5	Room temperature	—	Cu plate	Schneider et al. [22]
Ag-Ni film	$AgClO_4$: 0.01 M $Ni(ClO_4)_2$: 0.15 M $NaClO_4$: 0.1 M H_3BO_3: 0.15 M Thiourea: 0.2 M	3	—	—	Cu sheet	Liang et al. [23]
Ag-Ni film	$AgNO_3$: 0.01 M $Ni(NO_3)_2 \cdot 6H_2O$: 0.08 M H_3BO_3: 0.003 M Thiourea: 0.045 M	3.2	Room temperature	25	Cu plate	Srivastava et al. [24]
Ag-Ni solid film	$AgNO_3$: 1.69 g $Ni(NO_3)_2 \cdot 6H_2O$: 0.25 g H_3BO_3: 0.02 g in 100 mL distilled water	12	Room temperature	100	Cu plate	Srivastava et al. [25]
Ag-Ni nano particles	$AgNO_3$: 1.69 g $Ni(NO_3)_2 \cdot 6H_2O$: 0.25 g H_3BO_3: 0.03 g Thiourea: 0.35 g in 100 mL distilled water	3.6	80	20, 70, and 100	Cu plate	Srivastava et al. [26]

Metastable Ag-Ni alloy films are deposited by electrodeposition method by using their respective salts and some additives (such as thiourea, saccharin, trisodium citrate, citric acid, sodium perchlorate, etc.). The electrolyte is the electrochemical solution of metal or alloy salts that is required to deposit at cathode (conductive substrate) surface during electrolysis. The deionized and distilled waters are used as a solvent for most of the electrolytes; therefore, these electrolytes are called aqueous electrolytes. The costs of the aqueous electrolytes are lesser than the nonaqueous electrolytes due to the easy availability of the water solvent. Deionized water is water that has been treated to remove all the dissolved mineral salts. While, the distilled water is prepared by boiling the normal water so that it evaporates and then recondensed, leaving most impurities behind [31].

Mostly sulfate and nitrate salts with some suitable additives (such as thiourea, sac- charin, trisodium citrate, citric acid, sodium perchlorate, etc.) are used for the preparation of the electrolytes for deposition of the Ag-Ni systems. The sulfate and nitrate salts have high solubility in the water. The additives in the electrolyte act as grain refiners, complexing agents, brighteners, etc. The additive of sodium perchlorate is commonly used for reducing the resistance (i.e., viscosity) and increasing the ionic conductivity of the electrolytes [32]. And for improving quality of Ag in the deposited film, additives of boric acid and sodium gluconate are generally added in the electrolyte [33, 34]. List of some typical salts of metal ions with and without additives which are often used for deposition of Ag-Ni films/nanoparticles and the involved deposition process parameters are shown in **Table 1**.

Electrodeposition parameters for the Ag-Ni alloy films

The solid solubility, structure, and morphology of the deposited Ag-Ni alloy films depend on the several electrodeposition process parameters. Some of the crucial process parameters influencing the miscibility, morphology, and structure of the Ag-Ni alloy films are discussed in given below subsections.

Deposition current density

According to the electro-crystallization theory, whenever a high current density is applied, a high overpotential is generated at the cathode, as a consequence of this, the nucleation rate increases. The faster nucleation rate reduces the particle sizes of the deposited film [35–39], because the fine particle sizes increase the specific area or curvature that promotes the miscibility of the constituent elements [40, 41]. This phenomenon is known as the Gibbs-Thomson effect.

The composition of the Ag-Ni alloy film could be altered by changing the current density. Santhi et al. [32] have deposited metastable Ag-Ni film with the granular kind of morphology. The average particle size was ~50 nm. They observed that the Ni content in the deposited Ag-Ni film increases with increasing current densities. Bdikin et al. [42] reported that the particle size of the deposited Ag-Ni film decreases with increasing Ni contents in the film. For 70 wt% of Ni content, the Ag-Ni film particle size goes down to very small (~3.3 nm). Eom et al. [20] investigated the effect of the current density on the chemical composition and crystallography of the deposited Ag-Ni film. They deposited Ag-Ni films on a brass substrate using $NiSO_4$, $AgNO_3$, and $C_6H_5Na_3O_7$ electrolyte. The Ag content in the deposited Ag-Ni film was found to be higher at lower current density. The high content of Ag in the deposited film was due to its lower reduction potential. However, the content of Ni in the deposited films was found to be increased with increasing current density. At higher current density, Ni becomes the dominant component because of the more transfer limitation of Ag ions. Apart from the change in composition, the change in morphology as well as the grain size of the films with changing current densities was also reported by the authors. At a current density of 0.5 mA/cm², films had dendritic kind of morphology. But when the current densities were increased from 0.5 to 50 mA/cm², the morphologies were changed from dendritic to particle type (see **Figure 3**), and a significant reduction in grain sizes had also occurred.

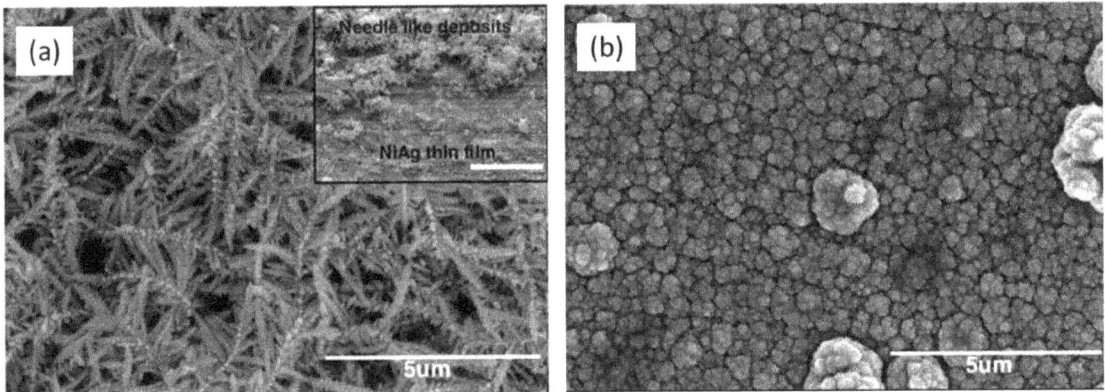

Figure 3. *SEM micrographs of Ag-Ni films deposited at current densities of (a) 0.5 mA/cm² and (b) 10 mA/cm² [20].*

The enhancement of miscibility of Ag and Ni elements to form a single-phase solid solutions of Ag-Ni alloy systems can be explained by Tafel equation (Eq. (4)) [43] and Eq. (5) [20] given below:

$$\eta = a + b \log i \tag{4}$$

$$J = K_1 \exp\left(\frac{-gs\varepsilon^2}{zekT\eta}\right) \tag{5}$$

where η is the overpotential at cathode, a and b are the constants, i is the current density, J is the nucleation rate, K_1 is the rate constant, g is the factor relating the surface area S of the nucleus to the perimeter P (g = P²/4S), s is the area occupied by one atom on the surface of the nucleus, T is the electrolytic temperature, k is the Boltzmann constant, and z, e, and ε are the number of electrons involved in the reaction, electron charge, and edge energy, respectively. According to Tafel equation (Eq. (4)), the reduction in the grain size is primarily due to the increase in the overpotential by increasing the current density. The relations between the nucleation rate, overpotential, and current density are given by Eqs. (4) and (5). From Eq. (5), it is clear that the nucleation rate has an exponential relationship with the overpotential. As the overpotential at the cathode increases, the nucleation rate also increases exponentially. In Eqs. (4) and (5), it can also be seen that the nucleation rate and current density are related to each other. Therefore, when current density increases, the overpotential at the cathode increases, as a result of this, the nucleation rate also increases. The increase in the nucleation rate produces finer grains in the deposited film. The finer grains increase the specific area or curvature which promotes an increase in the extent of miscibility of Ag and Ni atoms to form a single solid solution of Ag-Ni system.

Srivastava and Mundotiya [26] investigated the influence of the current density on the particles size of the deposited Ag-Ni alloy films. The Ag-Ni films were deposited on the Cu-substrate (30 mm X 10 mm X 0.3 mm) by using the electrolyte (0.17 g of $AgNO_3$, 0.25 g of $Ni(NO_3)_2 \cdot 6H_2O$, 0.03 g of H_3BO_3, and 0.35 g of thiourea in 100 ml of distilled water (pH ~ 3.6). During the electrodeposition, the temperature of the electrolyte was maintained at 80°C. The addition of an additive (complexing agent) of thiourea was chosen to deposit a granular kind of morphology in Ag-Ni film. They observed no significant change in morphologies of the deposited Ag-Ni films at current 20, 70, and 100 mA. At these current densities, the morphologies of the deposited films were similar granular morphology. However, the content of Ni in the Ag-Ni films increases, while the sizes of grains decrease with increasing current densities. These trends were in agreement with those reported in the scientific literature [20, 32, 42]. Thus, as a general statement, it can be concluded that the decrease in the grain size of the Ag-Ni film increases the miscibility of the constituent elements.

The miscibility behavior of the Ag-Ni system with different contents of Ni in Ag-Ni films can be explained by the X-ray diffraction (XRD) patterns shown in **Figure 4**. It can be seen in the patterns that the peaks of the Ag-Ni films shift to higher angles with increasing Ni content. The shifting of the peaks indicates substitution of the more Ag atoms by the relatively smaller Ni atoms with increasing content of Ni. That means the lattice constant of the Ag-Ni films decreases with increasing content of Ni. As the lattice constant decreases, the interplanar spacing of the Ag-Ni alloy films also decreases because the lattice constant is directly proportional to the interplane spacing. And the interplanar spacing is inversely proportional to the angle. Another important characteristic showed by the XRD profile is the broadening of the Ag-Ni films peaks with increasing Ni content. The broadening could be the result of decreasing grain size of the Ag-Ni films with increasing Ni content.

Therefore, the current density is the important parameter to control the miscibility of the Ag-Ni alloy system by controlling the size and composition.

Electrolyte temperature

The temperature of the electrolyte also influences the morphologies, the particles sizes, and the quality of the deposited single solid solution of the Ag-Ni films. This is due to the fact that the solubility of the metal salts in the electrolyte can be sufficiently increased by increasing the temperature of the electrolyte. As the temperature increases, the viscosity of the electrolyte decreases, and as a result of this, the mobility (diffusion and migration) of the metal ions toward the cathode increases [36]. Due to an increase in solubility and mobility of the metal ions in the electrolyte at a higher temperature, the conductivity of the electrolytes also improves, because the conductivity of the electrolyte depends on the degree of dissociation of dissolved metal ions in the electrolyte and the mobility of dissolved metal ions. The improvement in the conductivity of electrolyte at higher temperature decreases the overpotential at the cathode [44, 45]. A decrease in the cathodic overpotential with an increase in the electrolyte temperature above 55°C is reported by Rashidi and Amadeh [46]. The decrease in the cathodic overpotential slows down the nucleation rate; at low nucleation rate, the existing nuclei grow faster.

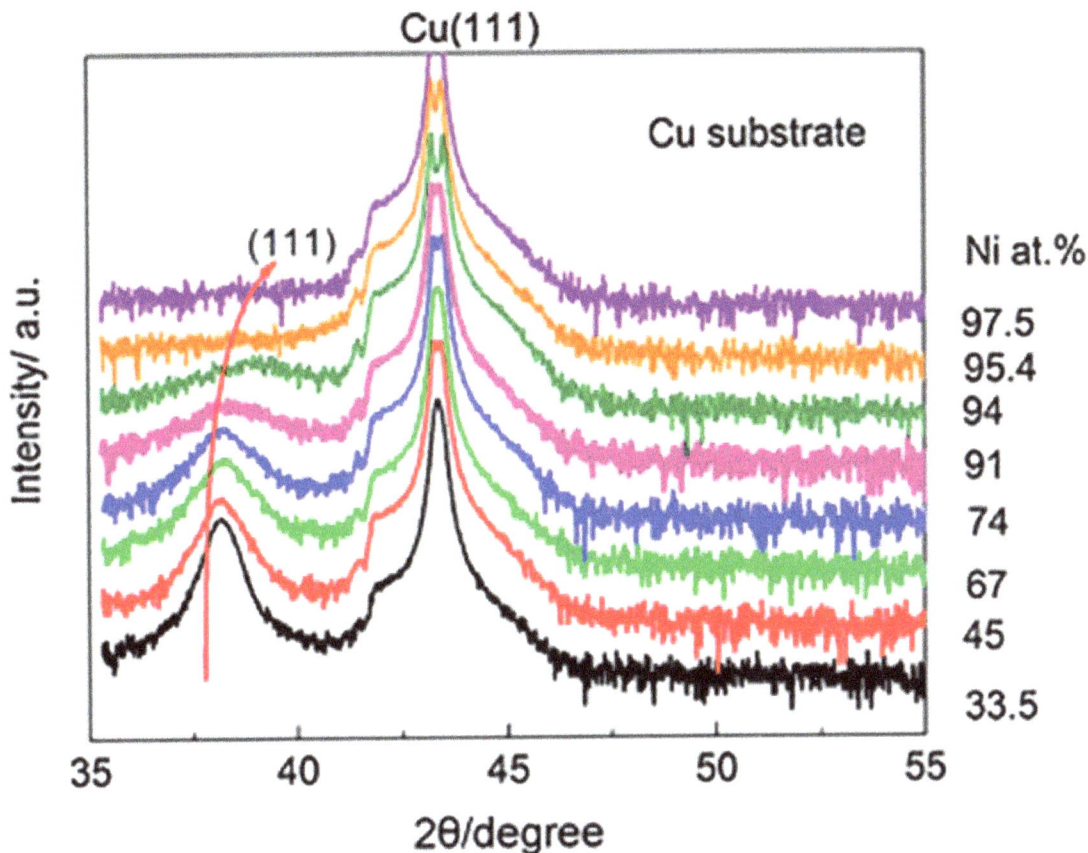

Figure 4. *XRD pattern of Ag-Ni alloy films with different Ni-content [23].*

Thus it can be concluded that an increase in the temperature of the electrolyte may lead to increased particle/grain growth in the deposited Ag-Ni film [26, 46, 47].

The temperature of the electrolyte also influences the composition of the deposited film. The concentration of Ag in the deposited film increases with increasing electrolyte temperature [48]. To deposit the films of the same compositions at different electrolyte temperatures, very high overpotential at the cathode is required. But high overpotential at the cathode favors hydrogen evolution which is undesirable for deposition of high-quality alloy film because it creates pores and/or hydrogen embrittlement in the deposit films [48].

Additives

The addition of the suitable additives in the electrolyte affects the morphology and the quality of the deposited film. The additives in the electrolyte produce an additional overpotential at the cathode [36] which helps in the creation of a homogeneous film with a low level of roughness [49]. For improving the quality of the deposited film, many suggestions have been given by the researchers. Among those suggestions, the most common suggestion is related to the blocking of lattice sites by adsorption of additives on the surface of the substrate. This helps to prevent the growth of grains in the deposited film by inhibiting short-range surface diffusion [50].

The reduction of Ag and Ni takes place at a potential of +0.2 and −0.26 V, respectively. The difference between the reduction potentials of Ag and Ni is too large. Due to the existence of this large difference in the reduction potentials of Ag and Ni during electrodeposition of Ag-Ni film, morphological instabilities between Ag and Ni phases occur during the growth process of the Ag-Ni film. Therefore, synthesizing a perfect Ag-Ni film by co-deposition of Ag and Ni is difficult in this condition. The deposited Ag-Ni film generally has a porous and dendritic morphology. For minimizing the difference of the reduction potentials of Ag and Ni, some suitable additives such as sodium citrate ($C_6H_5Na_3O_7$), thiourea, etc. are commonly added with electrolyte [20, 34]. The morphology of the electrodeposited Ag-Ni film can be tuned by the addition of suitable additives and/or complexing agents in the electrolyte.

These additives shift the reduction potentials of Ag and Ni more closely. The thiourea is one of the additives which has a strong complexing capacity with Ag [41] but no or weak complex with Ni. Due to its strong complexing capacity with Ag, it decreases the reduction potential of Ag into more negative potential without altering the reduction potential of Ni. The minimization of the difference of the reduction potentials of Ag and Ni by addition of additive in electrolyte also helps in suppressing the evolution of the hydrogen at the cathode. After the addition of the additive of thiourea in the electrolyte solution, a change of morphology of the feature of the deposited Ag-Ni films is observed [51]. The dendritic morphology changes to nanoparticles by adding additives in the electrolyte solution as shown in SEM micrographs in **Figure 5**. This transformation of dendritic morphology into nanoparticles after addition of thiourea additive in the electrolyte solution can be attributed to the minimization of the difference of the reduction potentials of the Ag and Ni elements.

The decrease of the reduction potential of Ag by adding additive of thiourea in the electrolytic solution is also reported by Liang et al. [23]. They deposited Ag-Ni films on the Cu substrate by using thiourea-based acidic solutions as an electrolyte. It was observed that the presence of thiourea in the electrolyte decreases the reduction potential of Ag via formation of the metal-organic complex. The average grain size in the deposited film was found to be lesser than 200 nm. It was also detected that the deposited films consist of a mixture of various phases such as a supersaturated solid solution of Ag-rich phase, pure Ni phase, and an amorphous phase at the boundaries. These phases were metastable. However, when the Ag-Ni alloy films were annealed at 600°C, these metastable phases were separated into Ag phase and Ni phase. Bdikin et al. [42] have also confirmed the separation of supersaturated solid solution (metastable phase) of Ag-Ni film into a single elemental phase after annealing the film at 600°C. They deposited the Ag-Ni films of different compositions on the copper substrates by using a single electrolyte containing Ag and Ni ions at room temperature and the current density 2–5.4 mA/ cm². The authors detected that the deposited nanocrystalline Ag-Ni films were metastable. These deposited Ag-Ni films become unstable with increasing annealing temperatures and decompose into a single elemental phase when annealed at 600°C.

Curvature of the deposited film's particles

The curvature provided by the particles of the deposited films is also playing a very crucial role in increasing the miscibility of the constituent elements.

Srivastava and Mundotiya [25] explored the possibility of changes in the extent of miscibility of Ag and Ni elements with changes in the system's morphology and tried to establish a correlation between size, morphology, and extent of miscibility in the Ag-Ni system. They deposited the Ag-Ni alloy films on the Cu substrate (30 mm × 10 mm × 0.3 mm) by using the electrolyte solution (1.69 g of $AgNO_3$, 0.25 g of $NiNO_3 \cdot 6H_2O$, and 0.02 g of H_3BO_3 in 100 mL of distilled water) at the current input of 100 mA and deposition time of 10 minutes. It was observed that the Ag-Ni films contain features of two distinct morphologies: (a) dendritic morphology (designated by type A feature) and (b) branched wire type morphology (designated by type B feature), as can be seen in **Figure 6(a)** and **(b)**, respectively.

The dendritic morphology (i.e., type A feature) had high Ag content in single- phase Ag-Ni when compared to the branched wire type morphology (i.e., type B of the feature). The SAED pattern from the type A feature shows a regular array of diffraction spots, which indicates that the dendritic morphology is a single crystal with size in micrometer range (**Figure 6(c)**), while the SAED pattern from the region of the type B feature consists of series of concentric rings, which indicates that the wire type morphology is polycrystalline in nature, and contains a large number of nanoparticles, and the rings in SAED pattern result in falling of the sum of individual patterns of these particles on the concentric rings (**Figure 6(d)**). The SAED patterns (both from the regions of the type A and type B features) did not show any diffraction for Ag, Ni, or even their oxides. The measured d-spacing values from the diffraction spots or concentric rings did not match with any of the d-spacing values for standard FCC crystal structures (i.e., Ag and Ni crystal structures). This indicates that the Ag and Ni atoms arranged themselves in solid solution atomic configurations to form a single solid solution of Ag-Ni system.

Figure 5. *SEM micrograph of Ag-Ni alloy film deposited at 100 mA/cm² and an electrolyte temperature of 80°C for: (a) without thiourea additive and (b) with thiourea additive in the electrolyte [51].*

Figure 6. *(a and b) TEM bright field image of type A and type B features. (c and d) SAED pattern obtained from type A and type B features [25].*

The average size of the nanoparticles in the branched wire type morphology was ~20 nm. It was observed that both features (i.e., type A and type B) contained Ag and Ni atoms in a single solid solution atomic arrangement, but the presence of Ni atoms in the feature made up of ~20-nm-sized particles (i.e., type B) was found to be larger than the type A feature. The greater extent of miscibility of immiscible Ag and Ni atoms in the type B feature can be attributed to the greater curvature provided by the nanoparticles.

Author details

Brij Mohan Mundotiya[1*], Wahdat Ullah[2] and Krishna Kumar[2]

1 National Institute of Technology (NIT), Hamirpur, Himachal Pradesh, India

2 Malaviya National Institute of Technology (MNIT), Jaipur, Rajasthan, India

*Address all correspondence to: brij2010iitrke@gmail.com

References

[1] Burton JJ, Hyman E, Fedak DG. Surface segregation in alloys. Journal of Catalysis. 1975;**37**:106-113. DOI: 10.1016/0021-9517(75)90138-4

[2] Zhu H, Yang S, Ni G, Yu D, Du Y. Fabrication and magnetic properties of $Co_{67}Ni_{33}$ alloy nanowire array. Scripta Materialia. 2001;**44**:2291-2295. DOI: 10.1016/S1359-6462(01)00761-8

[3] Yamasaki T. High-strength nanocrystalline Ni-W alloys produced by electrodeposition and their embrittlement behaviors during grain growth. Scripta Materialia. 2001;**44**:1497-1502. DOI: 10.1016/ S1359-6462(01)00720-5

[4] Lee W, Lee J, Bae JD, Byun CS, Kim DK. Syntheses of Ni_2Si, Ni_5Si_2, and NiSi by mechanical alloying. Scripta Materialia. 2001;**44**:97-103. DOI: 10.1016/ S1359-6462(00)00547-9

[5] Yichun L, Jiamin Z, Jikang Y, Jinghong D, Guoyou G, Jianhong Y. Direct electrodeposition of Fe-Ni alloy films on silicon substrate. Rare Metal Materials and Engineering. 2014;**43**:2966-2968. DOI: 10.1016/S1875-5372(15)60043-1

[6] Torabinejad V, Aliofkhazraei M, Assareh S, Allahyarzadeh MH, Rouhaghdam AS. Electrodeposition of Ni-Fe alloys, composites, and nano coatings—A review. Journal of Alloys and Compounds. 2017;**691**:841-859. DOI: 10.1016/j.jallcom.2016.08.329

[7] Guo H, Chen Y, Wen R, Yue GH, Peng DL. Facile synthesis of near-monodisperse Ag@Ni core-shell nanoparticles and their application for catalytic generation of hydrogen. Nanotechnology. 2011;**22**(19):195604. DOI: 10.1088/0957-4484/22/19/195604

[8] Rai RK, Srivastava C. Nonequilibrium microstructures for Ag-Ni nanowires. Microscopy and Microanalysis. 2015;**21**:491-497. DOI: 10.1017/ S1431927615000069

[9] Kabir L, Mandal AR, Mandal SK. Polymer stabilized Ni-Ag and N-Fe alloy nanoclusters: Structural and magnetic properties. Journal of Magnetism and Magnetic Materials. 2010;**322**:934-939. DOI: 10.1016/j.jmmm.2009.11.027

[10] Van Ingen RP, Fastenau RHJ, Mittemeijer EJ. Formation of crystalline Ag_xNi_{1-x} solid solutions of unusually high supersaturation by laser ablation deposition. Physical Review Letters. 1994;**72**:3116-3119. DOI: 10.1103/ PhysRevLett.72.3116

[11] Lee CC, Chen DH. Large-scale synthesis of Ni-Ag core-shell nanoparticles with magnetic, optical and anti-oxidation properties. Nanotechnology. 2006;**17**:3094-3099. DOI: 10.1088/0957-4484/17/13/002

[12] Tyler EH, Clinton JR, Luo HL. Electrical resistivity of metastable Ag-Ni alloys. Solid State Communications. 1973;**13**:1409-1411. DOI: 10.1016/0038-1098(73)90178-6

[13] Mundotiya BM, Srivastava C. Ag-Ni nanoparticles: Synthesis and phase stability. Electrochemical and Solid- State Letters. 2012;**15**(5):K41-K44. DOI: 10.1149/2.esl120008

[14] Singleton M, Nash P. The Ag-Ni (silver-nickel) system. Bulletin of Alloy Phase Diagrams. 1987;**8**(2):119-121. DOI: 10.1007/BF02873194

[15] Liu XJ, Gao F, Wang CP, Ishida K. Thermodynamic assessments of the Ag-Ni binary and Ag-Cu-Ni ternary systems. Journal of Electronic Materials. 2008;**37**:210-217. DOI: 10.1007/ s11664-007-0315-1

[16] Li ZC, Liu JB, Li ZF. Interface assisted formation of a metastable hcp phase by ion mixing in an immiscible Ag-Ni system. Journal of Physics: Condensed Matter. 2000;**12**:9231-9235. DOI: 10.1088/0953-8984/12/44/305

[17] Zhang Z, Nenoff TM, Huang JY, Berry DT, Provencio PP. Room temperature synthesis of thermally immiscible Ag-Ni nanoalloys. Journal of Physical Chemistry C. 2009;**113**:1155- 1159. DOI: 10.1021/jp8098413

[18] Kumar A, Damle C, Sastry M. Low temperature crystalline Ag-Ni alloy formation from silver and nickel nanoparticles entrapped in a fatty acid composite film. Applied Physics Letters. 2001;**79**:3314-3316. DOI: 10.1063/1.1414298

[19] Srivastava C, Chithra S, Malviya KD, Sinha SK, Chattopadhyay. Size dependent microstructure for Ag-Ni nanoparticles. Acta Materialia. 2011;**59**:6501-6509. DOI: 10.1016/j.actamat.2011.07.022

[20] Eom H, Jeon B, Kim D, Yoo B. Electrodeposition of silver-nickel thin films with a galvanostatic method. Materials Transactions. 2010;**51**(10):1842-1846. DOI: 10.2320/ matertrans.M2010126

[21] Raghupathy Y, Natarajan KA, Srivastava C. Anti-corrosive and anti- microbial properties of nanocrystalline Ni-Ag coatings. Materials Science and Engineering B. 2016;**206**:1-8. DOI: 10.1016/j.mseb.2016.01.005

[22] Schneider M, Krause A, Ruhnow M. Formation and structure of Ni-Ag layer by electrolytic deposition. Journal of Materials Science Letters. 2002;**21**:795- 797. DOI: 10.1023/A:1015710111396

[23] Liang D, Liu Z, Hilty RD, Zangari G. Electrodeposition of Ag-Ni films from thiourea complexing solutions. Electrochimica Acta. 2012;**82**:82-89. DOI: 10.1016/j.electacta.2012.04.100

[24] Srivastava C, Mundotiya BM. Electron microscopy of microstructural transformation in electrodeposited Ni-rich Ag-Ni film. Thin Solid Films. 2013;**539**:102-107. DOI: 10.1016/j. tsf.2013.05.080

[25] Srivastava C, Mundotiya BM. Morphology dependence of Ag-Ni solid solubility. Electrochemical and Solid- State Letters. 2012;**15**:K10-K15. DOI: 10.1149/2.003202es

[26] Srivastava C, Mundotiya BM. Size and solid solubility in electrodeposited Ag-Ni nanoparticles. Materials Science Forum. 2013;**736**:21-26. DOI: 10.4028/www.scientific.net/MSF.736.21

[27] Bona AD. Characterizing ceramics and the interfacial adhesion to resin: II—The relationship of surface treatment, bond strength, interfacial toughness and fractography. Journal of Applied Oral Science. 2005;**13**:101-109. DOI: 10.1590/S1678-77572005000200002

[28] Cao G. Nanostructures and Nanomaterials: Synthesis, Properties, and Applications. London: World Scientific; 2004. 175 p. DOI: 10.1142/9781860945960_0005

[29] Paunovic M, Schlesinger M, Snyder DD. Fundamental considerations. In: Schlesinger M, Paunovic M, editors. Modern Electroplating. 5th ed. New Jersey: Wiley; 2011. pp. 1-32. DOI: 10.1002/9780470602638

[30] Markov IV. Crystal Growth for Beginners: Fundamentals of Nucleation, Crystal Growth and Epitaxy. 2nd ed. Singapore: World Scientific; 1995. 422 p. DOI: 10.1142/9789812386243_0003

[31] US Water System. Deionized Water vs Distilled Water. 2018. Available from: https://www.uswatersystems.com/ deionized-water-vs-distilled-water. [Accessed: 2018-06-27]

[32] Santhi K, Karthick SN, Kim H, Nidhin M, Narayanan V, Stephen A. Microstructure analysis of the ferromagnetic Ag-Ni system synthesized by pulsed electrodeposition. Applied Surface Science. 2012;**258**:3126-3132. DOI: 10.1016/j. apsusc.2011.11.049

[33] Gomez E, Garcia-Torres J, Valles E. Study and preparation of silver electrodeposits at negative potentials. Journal of Electroanalytical Chemistry. 2006;**594**:89-95. DOI: 10.1016/j. jelechem.2006.05.030

[34] Kamel MM. Anomalous codeposition of Co-Ni: Alloys from gluconate baths. Journal of Applied Electrochemistry. 2007;**37**:483-489. DOI: 10.1007/s10800-006-9279-8

[35] Qu NS, Zhu D, Chan KC, Lei WN. Pulse electrodeposition of nanocrystalline nickel using ultra narrow pulse width and high peak current density. Surface and Coatings Technology. 2003;**168**:123-128. DOI: 10.1016/S0257-8972(03)00014-8

[36] Sharma A, Das S, Das K. Pulse electroplating of ultrafine grained tin coating. In: Aliofkhazraei M, editor. Electroplating of Nanostructures. Intech Open; 2015. pp. 105-129. DOI: 10.5772/61255

[37] Ibl N. Some theoretical aspects of pulse electrolysis. Surface Technology. 1980;**10**:81-104. DOI: 10.1016/0376-4583(80)90056-4

[38] Watanabe T. Nano-Plating Microstructure Control Theory of Plated Film and Data Base of Plated Film Microstructure. Oxford, UK: Elsevier Ltd; 2004

[39] Sharma A, Bhattacharya S, Das S, Das K. A study on the effect of pulse electrodeposition parameters on the morphology of pure tin coatings. Metallurgical and Materials Transactions A. 2014;**45**:4610-4622. DOI: 10.1007/s11661-014-2389-8

[40] Weissmueller J, Bunzel P, Wilde G. Two-phase equilibrium in small alloy particles. Scripta Materialia. 2004;**51**:813-818. DOI: 10.1016/j. scriptamat.2004.06.025

[41] Bellomo A, De Marco D, De Robertis A. Formation and thermodynamic properties of complexes of Ag(I) with thiourea as ligand. Talanta. 1973;**20**:1225-1228. DOI: 10.1016/0039-9140(73)80088-8

[42] Bdikin IK, Strukova GK, Strukov GV, Kedrov VV, Matveev DV, Zer'kov SA, et al. Growth, crystal structure and stability of Ag-Ni/Cu films. Materials Science Forum. 2006;**514-516**:1166-1170. DOI: 10.4028/ www. scientific.net/MSF.514-516.1166

[43] Burstein GT. A hundred years of Tafel's equation: 1905-2005. Corrosion Science. 2005;**47**:2858-2870. DOI: 10.1016/j.corsci.2005.07.002

[44] Tan AC. Tin and Solder Plating in the Semiconductor Industry: A Technical Guide. 1st ed. London: Chapman & Hall; 1993. 326 p. ISBN: 0442317123

[45] Inamdar AI, Mujawar SH, Barman SR, Bhosale PN, Patil PS. The effect of bath temperature on the electrodeposition of zinc oxide thin films via an acetate medium. Semiconductor Science and Technology. 2008;**23**:085013. (6pp). DOI: 10.1088/0268-1242/23/8/085013

[46] Rashidi AM, Amadeh A. Effect of electroplating parameters on microstructure of nanocrystalline nickel coatings. Journal of Material Science and Technology. 2010;**26**:82-86. DOI: 10.1016/S1005-0302(10)60013-8

[47] Sahaym U, Miller SL, Norton MG. Effect of plating temperature on Sn surface morphology. Materials Letters. 2010;**64**:1547-1550. DOI: 10.1016/j. matlet.2010.04.036

[48] Garcia-Torres J, Valles E, Gomez Influence of bath temperature and bath composition on Co-Ag electro-deposition. Electrochimica Acta. 2010;**55**:5760-5767. DOI: 10.1016/j. electacta.2010.05.014

[49] Gomez H, Lizama H, Suarez C, Valenzuela A. Effect of thiourea concentration on the electrochemical behavior of gold and copper electrodes in presence and absence of Cu(II) ions. Journal of the Chilean Chemical Society. 2009;**54**:439-444. DOI: 10.4067/ S0717-97072009000400026

[50] Schmidt WU, Alkire RC, Gewirth AA. Mechanic study of copper deposition onto gold surfaces by scaling and spectral analysis of in-situ atomic forces microscopic images. Journal of the Electrochemical Society. 1996;**143**:3122-3132. DOI: 10.1149/1.1837174

[51] Mundotiya BM. Phases and morphology in electrodeposited Ag-Ni film and nano-sized particles [thesis]. Indian Institute of Science Bangalore: Bangalore; 2012

2

Comparative Study and Simulation of Different Maximum Power Point Tracking (MPPT) Techniques using Fractional Control and Grey Wolf Optimizer for Grid Connected PV System with Battery

Mohamed Ahmed Ebrahim and R.G. Mohamed

Abstract

This chapter presents the comparative analysis between perturb & observe (P&O), incremental conductance (Inc Cond), and fractional open-circuit voltage (FOCV) algorithms using fractional order control & a new meta-heuristic called Grey Wolf optimizer (GWO) for extracting the maximum power from photovoltaic (PV) array. PV array systems are equipped with maximum power point tracking controllers (MPPTCs) to maximize the output power even in the case of rapid changes of the panel's temperature and irradiance. In this chapter, three cost effective MPPTCs are introduced: FOCV, P&O and Inc. Cond. The output voltage of the array is boosted up to a higher value so it can be interfaced to the local medium voltage distribution network.

Keywords: maximum power point tracking, grid connected photovoltaic, battery, grey wolf optimizer, boost converter, fractional order PI control

Introduction

Solar photovoltaic array system (SPVS) is one of the most prominent sources of electrical energy. SPVS is environmentally friendly and as a result there are no CO_2 emissions [1]. The energy dilemma represents in increasing the electricity production from the resources matching with the environment requires searching for new green, renewable, and sustainable ideas. SPVS along with wind turbines and fuel cells are possible innovative solutions for this dilemma [2, 3].

It was stated [4] that solar power capacity has expanded rapidly to 227 GW by the end of 2017 with a global growth rate of 26% which was higher than in 2016 (16.4%). Solar energy production was around 11% of the global renewable generation capacity, and increasing [4]. The installed capacity of SPVS in Egypt was about 1% of the total electricity production from renewable energy sources in March 2018 [5]. In SPVSs, the operation at the maximum power point (MPP) is necessity. As a result for this, various MPP tracking (MPPT) techniques are developed, investigated and implemented in the last decades [6]. One of the most powerful techniques is the fractional order PID (FOPID) based MPPT controller (MPPTC) [7]. These kinds of controllers have merged the merits of classical MPPTCs and PID controller [8]. However, FOPID based MPPTCs require efficient tuning methods to improve the dynamic response especially in the presence of system disturbances [9]. Thanks to Meta heuristic optimization techniques that can be employed to significantly tune MPPTCs.

In this chapter, design methodology for three different types of MPPTCs using fractional order PID (FOPID) is summarized.

This chapter also presents a new meta-heuristic called Grey Wolf Optimizer (GWO) inspired by grey wolves. The GWO algorithm mimics the leadership hierarchy and hunting mechanism of grey wolves in nature. Three main steps of hunting, searching for prey, encircling prey, and attacking victim, are implemented in this algorithm [10].

Practical case study

This study present PV solar power plant connected to the Egyptian national grid and installed in Kom Ombo, Aswan, Egypt. This power plant will have a total capacity of 20 MW which can be considered one of largest Egyptian PV project. The PV system is constructed using MATLAB/SIMULINK to mimic the actual system. Different scenarios are considered to test the effectiveness of the proposed MPPTCs. These scenarios include small as well as large environmental conditions changes.

Proposed system simulation

The simulation of grid-connected PV system with battery contains various simulation blocks such as PV array, battery, battery charge controller, three-phase voltage source inverter, the filter circuit, load, utility grid, and MPPT. **Figure 1** shows proposed system simulation. PV array is connected to the 11-kV network via a DC-DC boost converter and a three-phase three-level voltage source converter (VSC). In this paper, PV array generates a voltage of 666 V DC for a solar irradiance of 1000 W/m². The 100-kHz DC-DC boost converter is increasing voltage from PV natural voltage (666 V DC at maximum power) to 825 V DC. Switching duty cycle is optimized by an MPPT controller that using different techniques such as 'Incremental Conductance, Hill Climbing/Perturb and Observe (P&O), and Fractional Open-Circuit Voltage (VOC) techniques. This MPPT technique automatically varies the duty cycle to generate the required voltage to extract maximum power 1980-Hz 3-level 3-phase VSC. The VSC converts the 825 V DC link voltage to 300 VAC and keeps unity power factor.

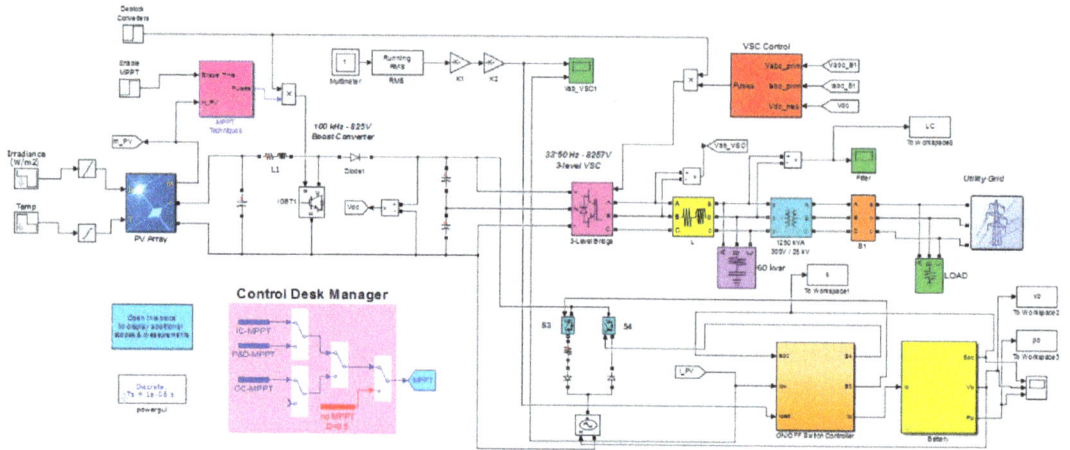

Figure 1. *Proposed system simulation.*

Problem overview

The most challenging problem considered by PV array system is how to auto- matically maintain the operation at maximum output power under environmental conditions continuous variation. In this chapter, a power converter that can vary the current coming from the PV array is employed as illustrated in **Figure 2** [6].

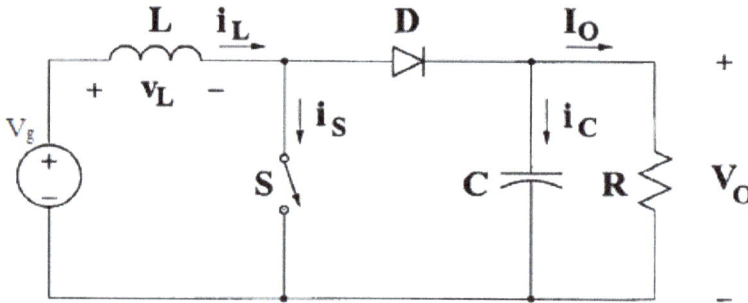

Figure 2. *Circuit diagram of boost converter.*

Figure 2 shows pulse width modulation based boost converter. The philosophy of operation of the converter depends on the on and off states of the switch S [11, 12]. The power converter (boost converter) parameters can be sized using the following equations [13]:

$$\frac{Vo}{V_g} = \frac{1}{1-D} \tag{1}$$

$$L = \left(\frac{V_g * D}{f * CRF} \right) \tag{2}$$

$$R_o = \frac{V_o}{I_o} \tag{3}$$

$$C = \frac{D}{(f \times R_o \times VRF)} \tag{4}$$

where D is the duty cycle ratio, Vg is the input voltage to boost converter, Vo is the output voltage from boost converter, f is the switching frequency, VRF is voltage ripple factor (according to IEC harmonics standard, VRF should be bounded within 5%), CRF is the current ripple factor (according to IEC harmonics standard, CRF should be bounded within 30%) [14] and R_o is the load resistance. We introduce the different MPPT techniques below in an arbitrary order.

Incremental conductance algorithm

Inc. Cond based MPPTC is derived from the fact that there are three operating regions around MPP. Each operating region has unique characteristics represented in the ratio between the power change and the voltage change. Roughly speaking, it can be considered that Inc. Cond based MPPTC is based on the slope of the PV array power curve [15, 16].

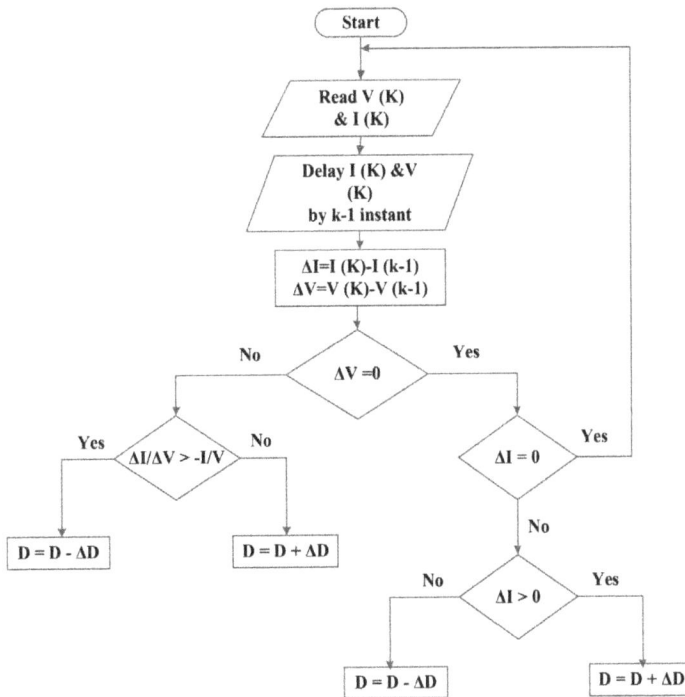

Figure 3. *Incremental conductance MPPT flowchart used for MATLAB simulation.*

$$\begin{cases} \dfrac{dP}{dV} = 0, \, at \, MPP \\[2mm] \dfrac{dP}{dV} > 0, \, left \, of \, MPP \\[2mm] \dfrac{dP}{dV} < 0, \, right \, of \, MPP \end{cases} \tag{5}$$

Since

$$\frac{dP}{dV} = \frac{d(IV)}{dV} = 1 + V\frac{dI}{dV} = 1 + V\frac{\Delta I}{\Delta V} \tag{6}$$

$$\begin{cases} \dfrac{\Delta I}{\Delta V} = -\dfrac{I}{V}, \text{ at } MPP \\[2mm] \dfrac{\Delta I}{\Delta V} > -\dfrac{I}{V}, \text{ left of } MPP \\[2mm] \dfrac{\Delta I}{\Delta V} < -\dfrac{I}{V}, \text{ right of } MPP \end{cases} \tag{7}$$

Thus, MPP can be tracked by comparing the instantaneous conductance (I/V) to the incremental conductance (ΔI/ΔV) as shown in the flowchart illustrated in **Figure 3**. The algorithm decrements or increments the duty cycle to track the new MPP. The increment size determines how fast the MPP is tracked.

Hill climbing/P&O algorithm

According to the sign of dP/dV where dP is the difference between power and dV is the difference between voltage of two succeeded point Hill climbing involves a perturbation in the duty ratio of the power converter [15–17]. The flow chart of the algorithm is shown in **Figure 4**. It is observed from P-V characteristic curve of the solar PV module that there are three main regions for operation. The first region is at the right hand side of MPP where the ratio between the power change over the voltage change (dP/dV) is negative. The second region is at the left hand side of MPP where the ratio dP/dV is positive. Moreover, the third region is at MPP exactly where the ration dP/dV is zero. P&O based MPPTC decides whether to increase or decrease the duty cycle depending on these three regions of operation.

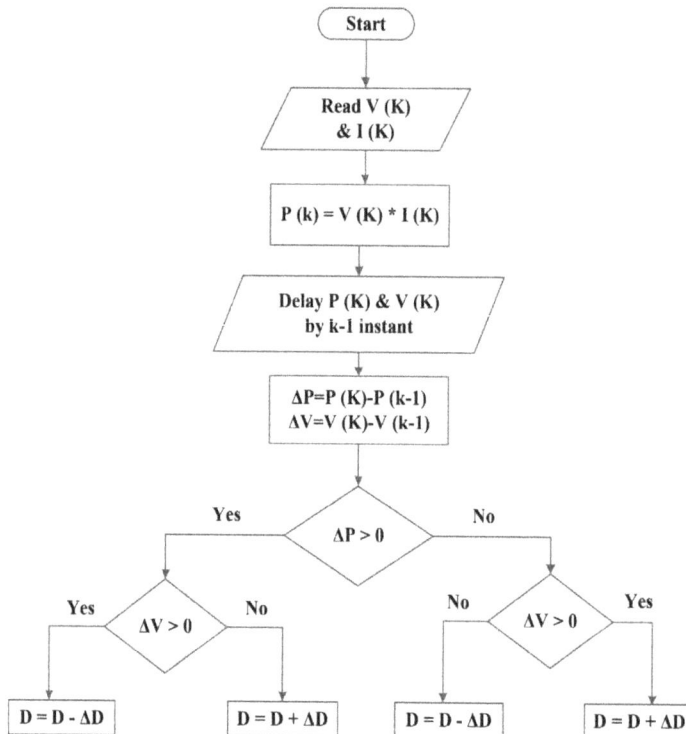

Figure 4. *Hill climbing/perturb and observe MPPT flowchart.*

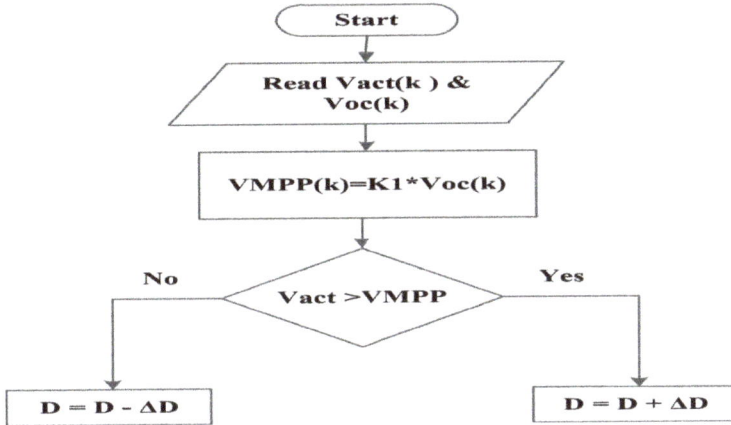

Figure 5. *Fractional open-circuit voltage algorithm.*

Fractional open-circuit voltage algorithm

The linear characteristic of V_{OC} under various operating conditions paves the way for FOCV based MPPTC [15, 18].

$$V_{MPP} \text{ ffi } K_1 \text{ x } V_{OC} \tag{8}$$

where K_1 is a constant of proportionality which depends on the characteristics of the PV panels. The algorithm of the fractional open circuit voltage is presented in **Figure 5**. The duty cycle is re-duced or increase by comparing V_{MPP} computed from V_{OC} and the actual voltage V_{act}. The factor K_1 ranges between 0.71 and 0.78.

Fractional order PID control

FOPID control is proven to provide more flexibility and ability to enhance modeling and control of systems' dynamics [19]. The transfer function of FOPID is given by

$$G(S) = K_P \left\{ 1 + \frac{1}{T_i s^\lambda} + T_d S^\mu \right\} \tag{9}$$

where K_p, T_i and T_d are controller gains while λ and μ are the integral and differential power in real number. By changing the values of λ and μ, the controller can be configured to behave within the four possibilities presented in **Figure 6** [20].

Figure 6. *Control space of FOPID.*

Figure 6 shows fractional PID control space. Recently, there are many optimization techniques are employed for solving engineering problems especially PI, PID, FOPI and FOPID based problems [19–39].

Grey wolf optimizer (GWO) technique

Grey wolves are considered as apex predators, meaning that they are at the top of the food chain [10]. **Figure 7** presents the social hierarchy of Grey wolves.

The mathematically model of the encircling behavior is represented by the following equations:

$$D = |CX_P - AX(t)| \qquad (10)$$

$$X(T + 1) = X_p(t) - AD \qquad (11)$$

The vectors A and C are calculated as follows:

$$A = 2A\, r_1 - a \qquad (12)$$

$$C = 2r_2 \qquad (13)$$

Note that the random vectors (r_1 and r_2) allow wolves to reach any position between the points illustrated in **Figure 8**. So a grey wolf can update its position inside the space around the prey in any random location by using Eqs. (10) and (11).

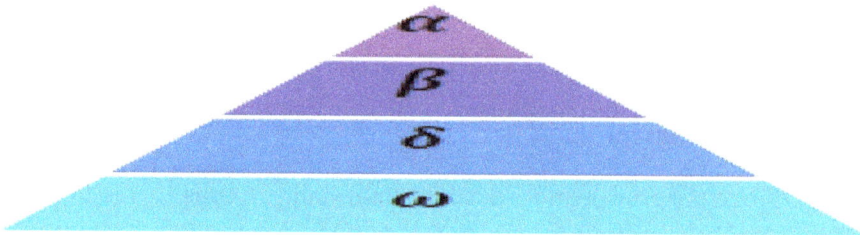

Figure 7. *Hierarchy of grey wolf (dominance decreases from top down).*

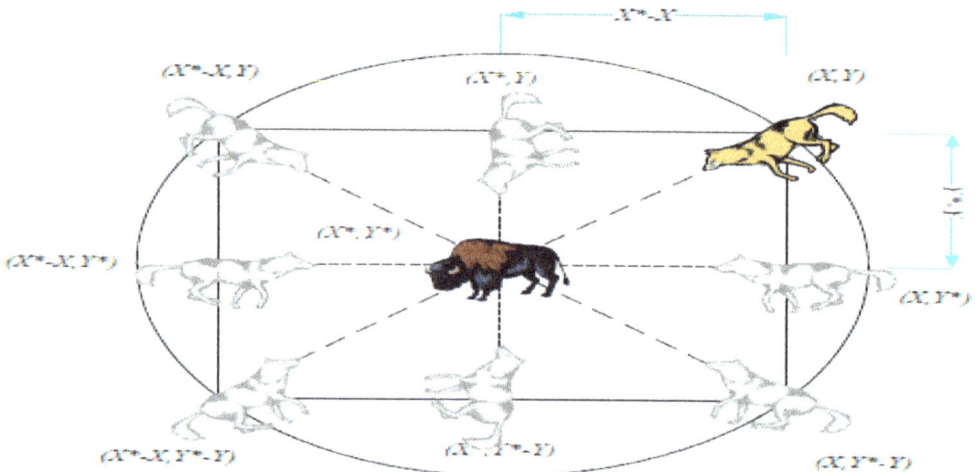

Figure 8. *Position vectors and their possible next locations.*

Simulation results and comparison

To validate the effectiveness of the proposed MPPTCs, the system under study quipped with only one MPPTC (P&O, Inc. Cond, FOCV, and FSCC) at a time is simulated. Wide range of operating temperature and irradiance is considered in this chapter to prove the superiority of GWO based MPPTCs over the conventional ones. The simulation results spot the light on the output voltage as well as power.

Perturb and observe method

The system equipped with P&O based MPPTC is simulated under small as well as large variations in temperature and irradiance. **Figure 9** demonstrates the dynamic response of the output voltage. The time response of the output voltage presents small voltage ripples during the rapid changes of temperature and irradiance. A proper filter can be employed to remove these ripples. In **Figure 10**, the dynamic time response for the output power is presented. The features of the time response for the system output power in case of P&O based MPPTC interprets that the P&O based MPPTC smoothly tracks MPP but with some oscillations especially at the transition intervals (high to low or low to high temperature and irradiance variations).

Incremental conductance method

The time response for the system voltage and output power is presented in **Figures 11** and **12** respectively. The dynamic response for the system output power in case of Inc. Cond based MPPTC is significantly improved compared to P&O case even in case of rapid variations in temperature and irradiance. **Figure 12** spotted the light on how Inc. Cond based MPPTC supersedes the P&O in smoothly tracking MPP.

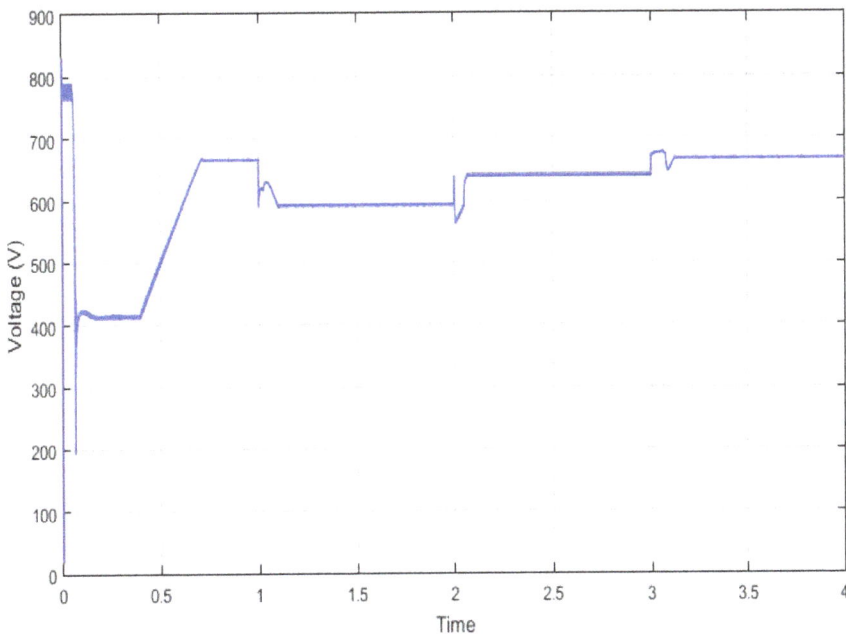

Figure 9. *P&O based MPPTC's output voltage waveform at different radiation and temperature.*

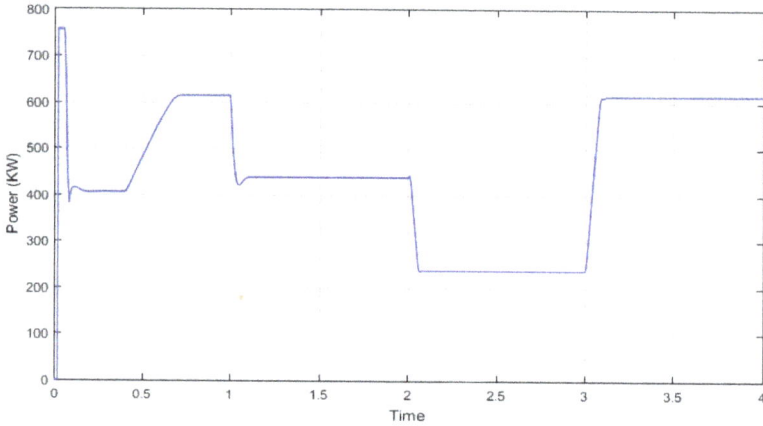

Figure 10. *P&O based MPPTC's output power waveform at different radiation and temperature.*

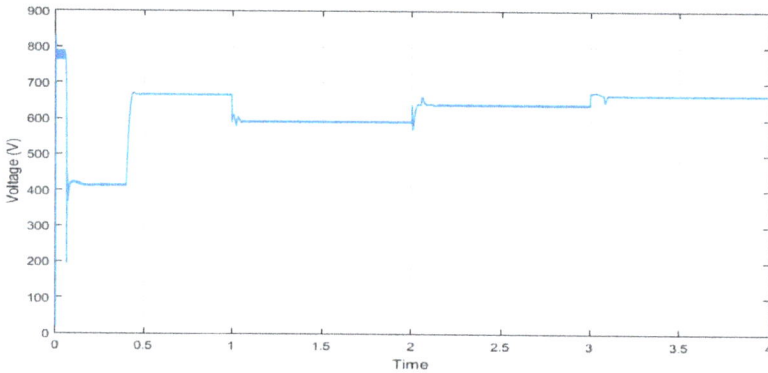

Figure 11. *Inc. Cond based MPPTC's output voltage waveform at different radiation and temperature.*

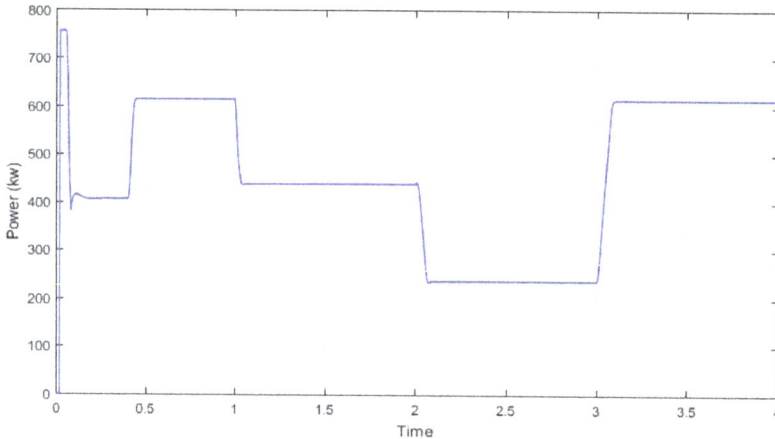

Figure 12. *Inc. Cond based MPPTC's output power waveform at different radiation and temperature.*

Open-circuit voltage method

The time response of the voltage and output power for the system equipped with FOCV based MPPTC is shown in **Figures 13** and **14** respectively. It is evident from the simulation results that the system response is poor especially in case of rapid changes in the operating conditions.

Table 1 presents a comparative study between the various applied MPPT techniques. It is worth mentioning that although Inc. Cond MPPT technique has good tracking response but it requires voltage and current measurements. Moreover, its implementation complexity is higher than P&O and fractional open circuit voltage methods.

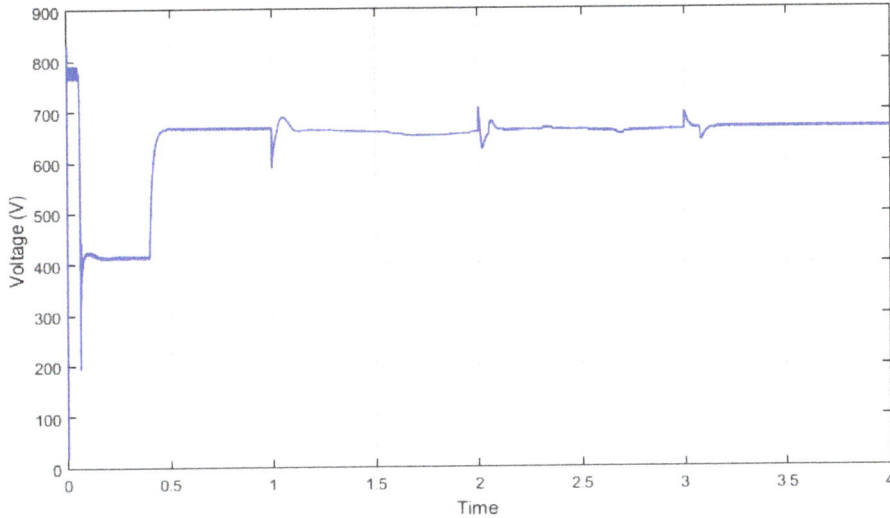

Figure 13. *FOCV based MPPTC's output voltage waveform at different radiation and temperature.*

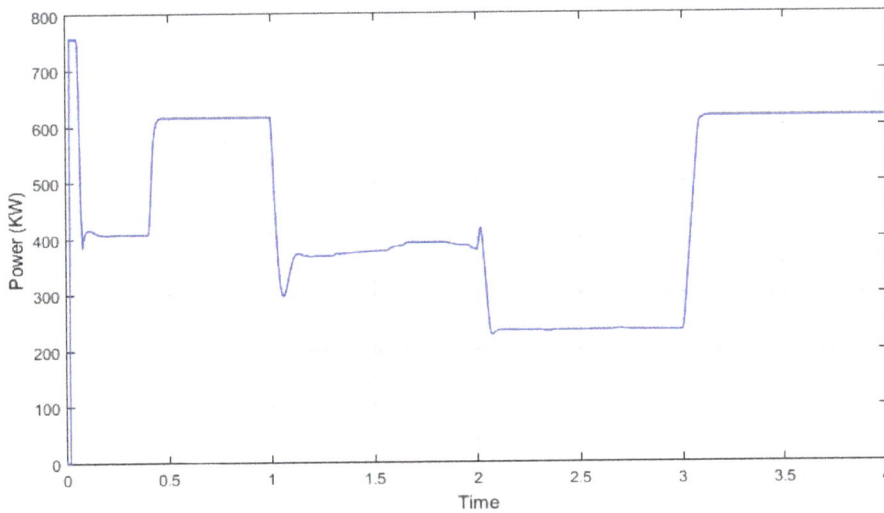

Figure 14. *FOCV based MPPTC's output power waveform at different radiation and temperature.*

Table 1. *Comparative study for various MPPT techniques.*

MPPT techniques	Parameters		
	Convergence speed	Implementation complexity	Sensed parameters
P & O method	Varies	Low	Voltage
Inc Cond method	Varies	Medium	Voltage, current
Fractional V_{oc} method	Medium	Low	Voltage

Conclusion

In this paper, four MPPT algorithms are implemented using the Boost converter. The models are simulated using MATLAB/SIMULINK. The simulation results show that P&O and Inc. Cond MPPTCs have better efficiency than FOCV and FSCC MPPTCs. Although, Inc. Cond provides good performance but its implementation has some challenges. Moreover, FOCV and FSCC based MPPTCs are very simple but both controllers lack to the accuracy due to their dependency on constant gains. Therefore, solar cell performance is significantly improved in the presence of MPPTCs. Hence, MPPTCs improvement has vital role in expanding the utilization of PV based systems.

Acknowledgements

The authors gratefully acknowledge the support of the Egyptian high education ministry, The Science and Technology Development Fund (STDF), and the French Institute in Egypt (IFE).

Conflict of interest

The authors of this chapter did not have 'conflict of interest' for publishing.

Author details

Mohamed Ahmed Ebrahim[1]* and R.G. Mohamed[2]

1 Faculty of Engineering at Shoubra, Benha University, Cairo, Egypt 2 Eastern Company, Giza, Egypt

*Address all correspondence to: mohamed.mohamed@feng.bu.edu.eg

References

[1] Wolfsegger C, Fraile D, Philbin P, Teske S. Solar generation-solar electricity for over one billion people and two million jobs by 2020. In: European Photovoltaic Industry Association, Renewable Energy House, Belgium and Greenpeace International, The Netherlands, Report No. 5. 2008

[2] Mohamed RG, Ibrahim DK, Youssef HK, Rakha HH. Optimal sizing and economic analysis of different configurations of photovoltaic systems. International Review of Electrical Engineering (I.R.E.E.). January–February 2014;9(1):146-156

[3] Mohamed RG, Ibrahim DK. Optimal sizing and economic analysis of stand- alone photovoltaic system. In: The 2012 World Congress on Power and Energy Engineering (WCPEE'12); Apr. Vol. 4. 2012. pp. 188-195

[4] Renewable Capacity Statistics 2018 (IRENA). https://www.irena.org/ publications/2018/Mar/Renewable-Capacity-Statistics-2018

[5] Annual Report 2018 (NREA). http:// www.nrea.gov.eg/Content/reports/ English%20AnnualReport.pdf

[6] Rekioua D, Achour AY, Rekioua T. Tracking power photovoltaic system with sliding mode control strategy. Renewable and Sustainable Energy Reviews. 2013:219-230

[7] Mukhopadhyay S, Chen YQ, Singh A, Edwards F. Fractional order plasma position control of the STOR-1M Tokamak. In: Joint 48th IEEE Conference on Decision and Control and 28th Chinese Control Conference. 2009. pp. 422-427

[8] Li Y, Chen Y, Ahn H-S. A generalized fractional-order iterative learning control. In: 50th IEEE Con-

ference on Decision and Control and European Control Conference (CDC-ECC). 2011. pp. 5356-5361

[9] Wang CY, Luo Y, Chen YQ. Fractional order proportional integral (FOPI) and [Proportional Integral] (FO [PI]) controller designs for first order plus time delay (FOPTD) systems. In: Chinese Control and Decision Conference. 2009. pp. 329-334

[10] Mirjalili S, Mirjalili SM, Lewis A. Advances in engineering software. Renewable and Sustainable Energy Reviews. 2014:46-61

[11] Khalid H, Mohamed TB, Ibrahim, Saad NB. Boost converter design with stable output voltage for wave energy. International Journal of Information Technology and Electrical Engineering. February 2013;2(1)

[12] Sathya P, Natarajan R. Design and implementation of 12V/24V closed loop boost converter for solar powered LED by lighting system. International Journal of Engineering and Technology (IJET). Feb-Mar 2013;5(1)

[13] Debashis DS, Pradhan SK. Modeling and Simulation of PV Array with Boost Converter. Rourkela: National Institute of Technology; 2011

[14] Hieu Nguyen X, Nguyen MP. Mathematical modeling of photovoltaic cell/module/arrays with tags in Matlab/ Simulink. Environmental System Research. 2015;1(1):1-13

[15] Esram T, Chapman PL. Comparison of photovoltaic array maximum power point tracking techniques. IEEE transactions on Energy Conversion. June 2007

[16] Stamatescua I, Făgărăşana I, Stamatescua G. Design and implementation of a solar-tracking algorithm. 24th DAAAM International Symposium on Intelligent Manufacturing and Automation, 2013. Procedia Engineering. 2014;1(1): 500-507

[17] Femia N, Giovanni Petrone M, Spagnuolo G, Vitelli M. Optimization of perturb and observe maximum power point tracking method. IEEE Transactions on Power Electronics. July 2005;20(4):963-973

[18] Sai Babu C. Design and analysis of open circuit voltage based maximum power point tracking for photovoltaic system. International Journal of Advances in Science and Technology. 2011;2(2):51-86

[19] Tajjudin M, Arshad NM, Adnan R. A design of fractional-order PI controller with error compensa-

tion. International Journal of Computer, Electrical, Automation, Control and Information Engineering. 2013;7(6):727-735

[20] Lachhab N, Svaricek F, Wobbe F, Rabba H. Fractional order PID controller (FOPID)-toolbox. In: European Control Conference (ECC) July 17–19, Zürich. 2013. pp. 3694-3699

[21] Aouchiche N, Aitcheikh MS, Becherif M, Ebrahim MA. AI-based global MPPT for partial shaded grid connected PV plant via MFO approach. Solar Energy. 2018;171:593-603

[22] Ebrahim MA, Becherif M, Abdelaziz AY. Dynamic performance enhancement for wind energy conversion system using moth-flame optimization based blade pitch controller. Sustainable Energy Technologies and Assessments. 2018;27: 206-212

[23] Benmouna A, Becherif M, Depernet D, Ebrahim MA. Novel energy management technique for hybrid electric vehicle via interconnection and damping assignment passivity based control. Renewable Energy. 2018;119: 116-128

[24] Maher M, Ebrahim MA, Mohamed EA, Mohamed A. Ant-lion inspired algorithm based optimal design of electric distribution networks. In: 2017 Nineteenth International Middle East Power Systems Conference (MEPCON); IEEE. 2017, December. pp. 613-618

[25] Aouchiche N, Cheikh MA, Becherif M, Ebrahim MA, Hadjarab A. Fuzzy logic approach based MPPT for the dynamic performance improvement for PV systems. In: 2017 5th International Conference on Electrical Engineering- Boumerdes (ICEE-B). October; IEEE. 2017. pp. 1-7

[26] Mohamed RG, Ebrahim MA, Bendary FM, Osman SAA. Transient stability enhancement for 20 MW PV power plant via incremental conductance controller. International Journal of System Dynamics Applications (IJSDA). 2017;6(4): 102-123

[27] Maher M, Ebrahim MA, Mohamed EA, Mohamed A. Ant-lion optimizer based optimal allocation of distributed generators in radial distribution networks. International Journal of Engineering and Information Systems. 2017;1(7):225-238

[28] Mousa ME, Ebrahim MA, Hassan MM. Optimal fractional order proportional—integral—differential controller for inverted pendulum with reduced order linear quadratic regulator. In: Fractional Order Con-

trol and Synchronization of Chaotic Systems. Cham: Springer; 2017. pp. 225-252

[29] Ebrahim MA, AbdelHadi HA, Mahmoud HM, Saied EM, Salama MM. Optimal design of MPPT controllers for grid connected photovoltaic array system. International Journal of Emerging Electric Power Systems. 2016; 17(5):511-517

[30] Mousa ME, Ebrahim MA, Hassan MM. Stabilizing and swinging-up the inverted pendulum using PI and PID controllers based on reduced linear quadratic regulator tuned by PSO. International Journal of System Dynamics Applications. 2015;4(4):52-69

[31] Jagatheesan K, Anand B, Ebrahim MA. Stochastic particle swarm optimization for tuning of PID controller in load frequency control of single area reheat thermal power system. International Journal of Energy and Power Engineering. 2014;8(2): 33-40

[32] Ebrahim MA, El-Metwally KA, Bendary FM, Mansour WM, Ramadan HS, Ortega R, et al. Optimization of proportional-integral-differential controller for wind power plant using particle swarm optimization technique. International Journal of Emerging Technologies in Science and Engineering. 2011

[33] Ahmed M, Ebrahim MA, Ramadan HS, Becherif M. Optimal genetic-sliding mode control of VSC-HVDC transmission systems. Energy Procedia. 2015;74:1048-1060

[34] Jagatheesan K, Anand B, Dey N, Ebrahim MA.

Design of proportional- integral-derivative controller using stochastic particle swarm optimization technique for single-area AGC including SMES and RFB units. In: Proceedings of the Second International Conference on Computer and Communication Technologies. New Delhi: Springer; 2016. pp. 299-309

[35] Ali AM, Ebrahim MA, Hassan MM. Automatic voltage generation control for two area power system based on particle swarm optimization. Indonesian Journal of Electrical Engineering and Computer Science. 2016;2(1):132-144

[36] Ebrahim MA, Elyan T, Wadie F, Abd-Allah MA. Optimal design of RC snubber circuit for mitigating transient overvoltage on VCB via hybrid FFT/ wavelet genetic approach. Electric Power Systems Research. 2017;143: 451-461

[37] Soued S, Ebrahim MA, Ramadan HS, Becherif M. Optimal blade pitch control for enhancing the dynamic performance of wind power plants via metaheuristic optimizers. IET Electric Power Applications. 2017;11(8): 1432-1440

[38] Ebrahim MA. Towards robust non- fragile control in wind energy engineering. Indonesian Journal of Electrical Engineering and Computer Science. 2017;7(1):29-42

[39] Ebrahim MA, Ramadan HS. Interarea power system oscillations damping via AI-based referential integrity variable-structure control. International Journal of Emerging Electric Power Systems. 2016;17(5): 497-509

3

Introductory Chapter: Electric Power Conversion

Marian Găiceanu

Chapter overview

The introductory chapter has in view an incursion in discovering electricity, how can be handled, and the future of it. The chapter starts with the ancient discovery of electricity. Starting from the Kite experiment to the energy use of lightning is mentioned in the *New Discoveries in the Electricity* section. Moreover, the current path from the electrostatic machine to ion wind propulsion system is mentioned in the same section. A short history of energy conversion technology is described in the forthcoming section. Different electric power conversion technologies are mentioned. The possible pathways of the future electric power conversion are mentioned. Some ideas about electric power conversion development are mentioned at the end of this chapter.

The ancient discovery of the electricity

Electricity is a natural phenomenon. One of the first known discovered natural electricity generations is the *electric fish* (discovered in twenty eighth century BC by the Egyptian). These types of fish are named *electrogenic* (one of the most dangerous fish is the *electric eel—electrophorus electricus*—due to the lethal electric shocks generation), being capable to generate the electric fields through electric organ discharge (EOD). Nowadays, the *bioelectrogenesis* (generation of the electricity by the living organisms) phenomenon is studied by the *electrophysiology* science branch. The receiver fish of the electric field is named *electroreceptive*, having the ability to receive the electric field through the *electroreception* feature [1].

The Thales of Miletus (around 600 BC) remarked the attraction of lightweight material by rubbed amber (ἤλεκτρον in Greeks or ēlektron), discovering static electricity [2]. However, as the archeological discoveries are on-going the researchers make hypothesis of electric light existence in ancient Egypt [3–6].

New discoveries in the electricity

In 1600 the scientist William Gilbert published the on the *Magnet treatise*, becoming the *Father of Electricity* [7]. In his published work *Pseudodoxia Epidemica* (1646), second book—*Tenets Concerning Mineral and Vegetable Bodies*, Thomas Browne used the word *Electricity*. The inventor of the *Electricity* (1663) was declared Otto von Guericke by producing static electricity by friction of a sulfur ball [8].

The first discover of conductors and insulators were made by the Stephen Gray (1731). Charles François de Cisternay du Fay discovered the existence of two electricity types: vitreous and resinous (1733), which were renamed positive and negative electricity by Benjamin Franklin in 1750. Through the bolt of lightning Benjamin Franklin discovers the connection between lightning and electricity, as observed in the *Kite Experiment* (1752) [9]. The bolt of lightning can be intra-cloud (**Figure 1a**), between clouds (**Figure 1b**) or between cloud and ground (**Figure 1c**) [10]. A massive steel sculpture was performed by Isamu Noguchi in Philadelphia [11], as a memorial to Benjamin Franklin (**Figure 1d**) [12]. An average value of the energy released during thunderstorm is about 10,000,000 kilowatt-hours (3.6×1013 joule) [13].

Taking into account such huge quantity of energy provided in a very short time, the scientists propose the use of the captured lightning energy as alternative energy in different kinds of patents [14, 15] in order to use the stored energy lately. In the future, as the nanotechnology penetrates the scientific knowledge environment the lightning energy could be a real alternative energy solution. In 1741 William Watson used a vacuum glow-lamp supplied by an electrostatic machine. Recently, at the end of 2018 the MIT engineers have been discovered the first no moving parts aircraft by using the electrostatic energy (**Figure 2**) [16]. They used the electro aerodynamic thrust principle to design ion wind propulsion system.

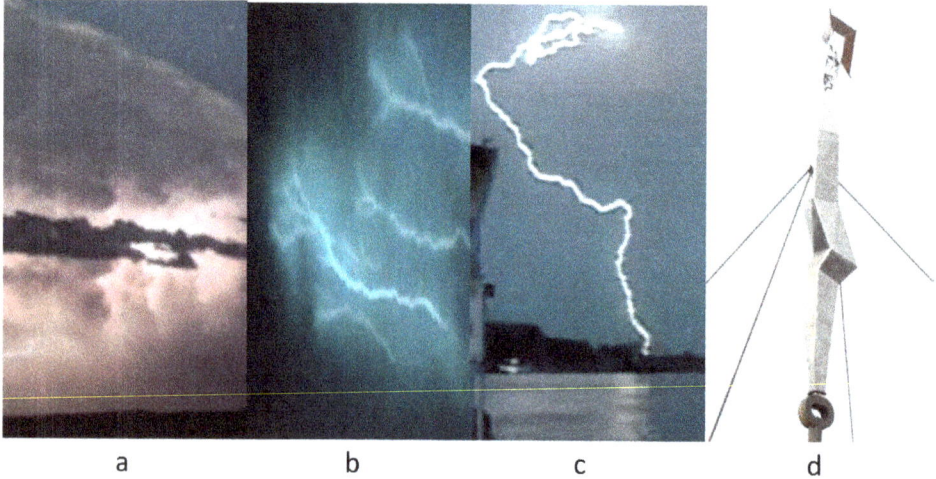

a   b   c   d

Figure 1. *The bolt of lightning (a) intracloud (b) between vertical clouds (c) between cloud and ground (d) bolt of lightning.*

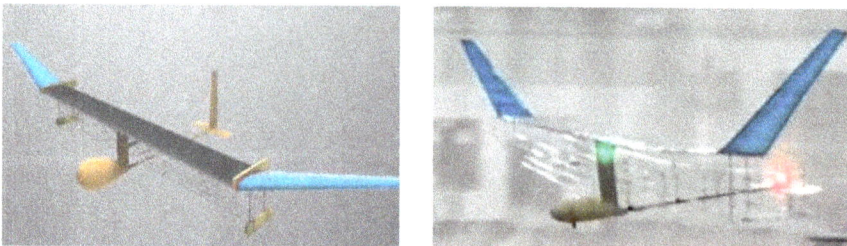

Figure 2. *The first no moving parts aircraft by using the electrostatic energy [17].*

The bioelectricity was discovered by Luigi Galvani in 1780. The voltaic pile was made in 1800 by Alessandro Volta (the international unit for electric potential difference—voltage—is volt). This invention was a huge step regarding the source of the electricity, being more reliable than the electrostatic generator.

The actual technologies in electrical energy storage could be reviewed in [18]. The NAWA Technologies introduced in mass production the new technology as ultra fast carbon battery which is carbon-based ultracapacitor with vertically aligned carbon nanotube (VACNT) [19].

The future of the energy storage could be graphene-based supercapacitors. This new type delivers four times more energy density than the current supercapacitors, maintaining high power and long operational life (Meilin Liu) [20]. In 1802, Humphry Davy invented the incandescent lamp.

A short history of energy conversion technology

The energy conversion technology starts with biomass (primarily wood: dried plants) conversion to heat. The mechanical energy was obtained to replace the animal or human power. In order to obtain the mechanical energy as primary energy sources wind (air flowing by using windmills) or flowing water (by using water- wheels) were used. The first notes of windmills were from Hero of Alexandria, in the first century of common era (CE). The ocean tides were the primary source of energy for tidal mills since 1086. The idea of conversion steam power to move a piston into cylinder (earlier pump) was delivered by Denis Papin (1679). The conversion of the thermal energy (steam) into hydraulic energy was patented by Thomas Savery (1698). Thomas Newcome, together with Savery invented the first piston-operated steam pump (1712). James Watt was the Father of the Watt steam engine (1765), a modern way of the steam engine.

The first steam powered locomotive was provided by Richard Trevithick (1803) [21]. Robert Stirling patented in 1816 the engine without high-pressure boiler (for safety reason). The mechanical energy conversion into electricity was discovered by Michael Faraday by using electrical generator (1830s).

Energy conversion technologies

The total energy of one system, according to the first thermodynamics law, is composed by heat and capacity to do the work. The standard unit of energy is *joule* (after the name of scientist James Prescott Joule). The efficiency of the energy conversion is defined by the second law of thermodynamics [22]. The energy cannot be created or destroyed, only transformed it from one form to another (law of conservation of energy). In other words, the energy within a reference frame the energy is always the same.

The energy per unit time is power. The unit *watt* was introduced by Carl William Siemens (1882) for *real power* in AC system. The *watt* name is provided after the James Watt.

According to [23] (the *metric EU directive*), the symbol of the *reactive power* is "*var*" and it was proposed in 1929 by the Prof. Constantin Budeanu and introduced in 1930 by the International Electrotechnical Commission in Stockholm, as international unit for reactive power [24].

There are two main states of energy potential (store) and kinetic (in motion, working). There are several forms of energy: chemical, nuclear, gravitational, elastic (stored), electrical, electromagnetic (radiant, light), mechanical, thermal (heat), ionization, sound. The energy conversion takes place through power converters. Power converters can be rotating (electromechanical) or static.

Electromechanical power converters

In 1821 Michael Faraday invented the electric motor (a primitive version). The relationship between the current and the voltage, known as Ohm's law, was quantified by Georg Ohm in 1827. Four years later, in 1831 Michael Faraday discovered the electromagnetic induction and the electromechanical conversion of the energy. Through the experiments, Faraday demonstrated that the electricity is obtained by friction, electromagnetic induction, chemical or thermoelectric. Werner von Siemens demonstrated the obtaining of the electric light from dynamo generators (1867). Siemens laid the groundwork of the modern electric generators. In treatise on electricity and magnetism (1873), James Clerk Maxwell published the unified theory of electricity and magnetism through Maxwell's equations.

Thomas Alva Edison found in 1870 the electric lighting solution by using ncandescent light bulb. In 1882 Edison opened the Electric Light Company, the first public electric power plant to supply 110 volts direct current (DC) for the lights. In Westinghouse the first hydroelectric power plant opened in 1882. Nikola Tesla invented the Tesla coil in 1883. In order to transmit the electric power through this type of transformer the low voltage is changed to high voltage. An induction motor supplied from the alternating current was developed by Tesla in 1887.

George Westinghouse developed the first multiple-voltage alternative current (AC) power distribution system by using the transformers, in 1886. The electric energy was produced from the hydroelectric generator. Westinghouse imported the transformers and Siemens AC alternators (generators) from Europe and formed Westinghouse Electric & Manufacturing Company in 1886, a power distribution plant. Galileo Ferraris was the "Father of three-phase current" (1885). In 1888, Galileo Ferraris published the work on the AC polyphase motor. In 1889, Mikhail Dolivo-Dobrovolsky invented the cage version of the three- phase induction motor and in 1890 the wound rotor version of the actual three- phase induction motor.

Nowadays, the power transmission over long distance is reliable in AC power system. It requires high voltage obtained by using transformers conducting to low currents; therefore, low transmission losses occur. However, by introducing the static power converters, the DC transmission becomes a challenge. The generation of the electricity mostly takes place into a power station by using the electromechanical generators. The primary sources of the generators are heat engines fueled by fossil sources or nuclear fission but also by renewable energy: wind, water motion energy, solar or geothermal power.

Static power converters

The development of power electronics starts with French scientist J. Jasmin. Jasmin discovered in 1882 that the mercury electric arc assures conduction in one way [25]. By using this property, the rectifier had been born transforming alternating current (AC) in direct current (DC). In 1892, L. Arons invented the vacuum valve with mercury arc. In 1906, J. A. Fleming invented the vacuum diode. The silicon valve was invented by G. W. Pickard in 1906. The vacuum triode was invented by L. de Forest in 1907. The first production of ignitrons was made by Westinghouse Company in 1933. Starting with the transistor invention in 1948, the next generation of power converters arises. The silicon controlled rectifier (SCR) or thyristor, was invented in 1956 (John Moll). During 1956–1975, the power converters based on SCRs had been developed.

The new generation (1975–1990) of the power converters starts with new invented power switches: MOSFETs (metaloxide-semiconductor field-effect transistor, 1980), bipolar transistors and power bipolar junction transistors (BJT), the power gate turn-off (GTO) thyristors. The next generation of the power converters starts with the microprocessors, and application specified integral circuits (ASIC). The automatic control of power converters improves significantly the performances. The introduction of the intelligent power modules (IPM) boosts the performances of the power converters.

The main types of power converters are: rectifier, dc-dc chopper, inverters, direct (cycloconverters and matrix converters) and indirect (back-to-back) frequency converters. From commutation point of view, the power converters are natural commutated and force commutated. The phase-controlled rectifiers are natural commutated and used at high power. Due to the power quality problems, the performances are lower than of the force commutated power converters. Besides of the development of the power devices and of the drivers, the modulation strategy is a key factor of efficiency improvements allowing the increased power capacity. The most common modulation strategy is pulse width modulation (PWM).

In order to obtain a variable speed, the alternating machines (synchronous and asynchronous) are combined with power converters. The most advanced control methods of the alternating machine are scalar control, vector control (field oriented), direct torque control, and sensorless control. The general exploration of the man-machine visual interactivity in the sense of the metaheuristic search algorithms can be a modern way for the energy domain exploitation. The modern algorithms based on artificial intelligence to find the optimal solution are also applied in this stringent energy area. The features of the modern drives include remote control, networking, human- machine interface functions [26]. The new power converters should be protected against threats due to the digital communication and control. Therefore, the cybersecurity for the smart power converters should be taken into account [27]. Artificial intelligence and machine learning algorithms will be part of modern drive control systems.

Overview of the of the energy system

By knowing the input (energy production) and the output (energy consumption) of the the energy system the adequate mathematical model can be deducted. Taking into account the energy production (**Figure 3**—net electricity generation was 3.10 million GWh in 2016 [28]) from primary sources and the final energy consumption (**Figure 4**—EU 2016 energy consumption 2.78 million GWh) the main *energy conversion technologies* can be described briefly.

The hot water from the geothermal natural sources is used in steam power plant to obtain the electricity. The kinetic and potential energy from the rivers are trans- formed in electric energy through the hydro power station. The windmill captures the power of air flow to obtain the mechanical power. The mechanical power is converted into electricity by using the electric generators. The nuclear power plant used the nuclear energy to obtain the electricity. The photovoltaic cells are used to convert the solar energy into electricity. The most pollutant power conversion technologies are based on the fossil fuels. The transport sector is the most dominant consumption one where the fossil fuels are dominant, followed by the industry sector and households (**Figure 4**) [29]. Therefore, alternative energy solutions as environmentally friendly way should be taken into account. The prediction of the world energy consumption between 2010 and 2040 is an increase by 56% [30].

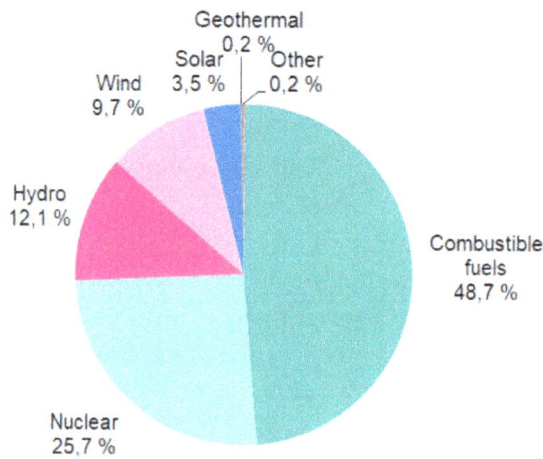

Figure 3. *Energy production [28] (% of total, based on GWh), EU-28, 2016.*

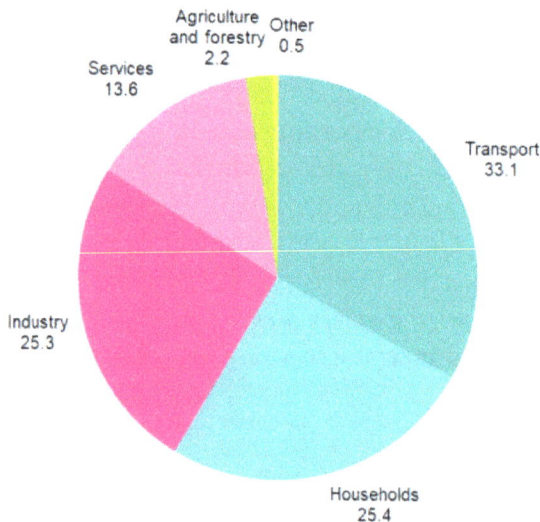

Figure 4. *Energy consumption [29] (% of total, based on GWh), EU-28, 2016.*

Currently, almost the entire world energy production is covered by hydropower plants, thermal power stations, and nuclear power plants. The environmentally friendly renewable energy will expand in the future.

The first optical amplification of the light based on stimulated emission of electromagnetic radiation, known as light amplification by stimulated emission of radiation (LASER) was invented in 1960 by Theodore H Maiman. The application of laser could be in energy transmission between the space stations and earth or could be used for the propulsion of the spatial vehicle [31]. Recently (2018), the Lawrence Livermore National Laboratory (LLNL) has obtained record of 2.15 megajoules (MJ) of energy with the National Ignition Facility (NIF) laser system [32]. The main purpose of NIF is to produce "green" electric power by fusion without producing radioactive waste. Nobel prize in physics 2014 was awarded for the invention of the blue light-emitting diodes (LEDs), helped to create energy-efficient light sources in a completely new way. The impact of discovering white LED lamps will decrease significantly the world's electricity used for lighting [33].

The extreme light infrastructure—nuclear physics (ELI-NP) pan European project has the main outcome the highest intensity laser system consisting of two 10 PW laser arms. The laser system will develop laser beam of 10^{23} W/cm^2 power intensity and up to 10^{15} V/m electrical fields [34]. The current record is HERCULES Petawatt laser at the University of Michigan, USA. The focused laser beam has the intensity of 2×10^{22} W/cm^2 [35, 36]. For a comparison, the strength intensity of the lightning electric field is 3×10^6 V/m (30 kV/cm) under dry air condition, at atmo- spheric pressure [37, 38]. The methods of measuring of these strength fields are described in [39]. The project ELI-NP was promoted by Professor Gérard Mourou in 2006 [40], one of the winners of the Nobel prize in physics 2018 [41]. By using laser technology, laser power transfer will be a solution to power the space vehicles or interplanetary communications.

Cyber physical systems

In order to increase the safety, security and reliability of the electricity the entire systems of energy domain should integrate the modern cyber-security features. Cyber physical systems (CPS) are based on the special architecture [42]. Two key enabled factors conduct to the design of modern energy plant infrastructures:

1. due to the limited flexibility and vulnerability of the ancient bulk centralized power systems;

2. protecting the bidirectional (receiving the adequate signals from the devices, transmitting the adequate control signals) communication data.

According to [43], there are two main causes that enable major changes to the electricity infrastructure:

- Environmental sustainability

- Effective management of pervasive data and extracted information

Two deterministic CPS models are described in [44], and used for practical realization of the distributed CSP through two projects: PRET (precision timed (PRET) computation in cyber-physical systems) [45], and Ptides (programming temporally-integrated distributed embedded systems) [46].

The pathways of the cyber physical systems developing should conduct to open CPS (cyber-physical systems) platforms [47]. The vision or short-term research strategy of any country should include the integration into Continental Cyber Physical Platforms (CCPP), respectively Intercontinental Cyber Physical Platforms (ICCPP). One development direction already started through the Open Source Cyber Platform [47]. At the same time, one of the European initiatives is made by the European Strategy Forum on Research Infrastructures [48].

Some of the main CPS design features could be mentioned as follows: to be modular, easily interconnecting with other CPS (plug and play), communication, self-healing, restoration, resiliency, autonomy, robustness, sensor-connectivity, and security assurance. In order to ensure world protection against internal or external attacks, the syn- ergies between the local CPS should be enabled, integrating in one intercontinental platform (ICCPP). One of the European projects has as main objective to disperse the CPSs to market through the seven platforms [49]. According to [50], the information about the current continental research platform in Europe is available. The European research infrastructures in Energy area could be finding on [51].

Conclusion

As the electric energy has the great importance for human development, in order to obtain it, any research strategy should include the electric power conversion development. Nowadays, the symbiosis between energy and information and communications technology is already taken into consideration. However, one key success of fast technology developing is the sharing of knowledge through open source. All the budget funded research projects should provide back the accumulated research results to the entire world. A win-to-win mechanism to access the high-tech laboratories by the worldwide researchers has been already created at least by the European Parliament and US officials. Another key-enabled of fast technology developing is the adequate training of the human resources through the advanced and innovative teaching, and researching methods. The curriculum development and adequate tools for education should be adapted very fast from the Digital Era [52] to the forthcoming Cyborg Era [53].

Acknowledgements

This work was supported by a grant of the Romanian National Authority for Scientific Research, CNDI—UEFISCDI, project number PN-II-PT-PCCA-2011-3.2-1680.

Author details

Marian Găiceanu[1,2]

1 Dunarea de Jos University of Galati, Galati, Romania

2 Integrated Energy Conversion Systems and Advanced Control of Complex Processes Research Center, Galati, Romania

*Address all correspondence to: marian.gaiceanu@ieee.org

References

[1] Benedict K, Yuzhi H, Long John A. Electroreception in early vertebrates: Survey, evidence and new information. Palaeontology. 2018;**61**(3):325-358. DOI: 10.1111/pala.12346

[2] Keyser PT. The purpose of the parthian galvanic cells: A first century A.D. electric battery used for analgesia. Journal of Near Eastern Studies. 1993;**52**(2):81-83

[3] The Library of Congress. Everyday Misteries [Internet]. 2017. Available from: http://www.loc.gov/rr/scitech/ mysteries/static.html [Accessed: January 10, 2019]

[4] Ancient Code, Ivan Petricevic. Did Ancient Egyptians Have Electricity? 2012. Available from: https://www. ancient-code.com/did-ancient- egyptians-have-electricity [Accessed: January 10, 2019]

[5] Larry Brian Radka. A Short History of Ancient Electricity. 2006. Available from: https://www.bibliotecapleyades. net/ciencia/ciencia_hitech05.htm [Accessed: January 10, 2019]

[6] Frank Dörnenburg, "Electric Lights in Egypt?" Available from: http:// WorldMysteries.com, http:// old.world- mysteries.com/sar_lights_fd1.htm [Accessed: January 10, 2019]

[7] Available from: https://www. smashinglists. com/10-first-electricity- milestones/ [Accessed: January 10, 2019]

[8] Guarnieri M. Electricity in the age of enlightenment. IEEE Industrial Electronics Magazine. 2014;**8**(3):60-63. DOI: 10.1109/MIE.2014.2335431

[9] National Archives, The Kite Experiment; 19 October 1752 [Internet]. 2017.Available from: https:// founders.archives.gov/documents/Franklin/01-04-02-0135 [Accessed: January 10, 2019]

[10] Lightning and Thunder. Encyclopedia Britannica [Internet]. Available from: https://www.britannica. com/science/lightning-meteorology/ media/340767/194043 [Accessed: January 10, 2019]

[11] The Electric Ben Franklin [Internet]. 1995. Available from: http://www. ushistory.org/franklin/philadelphia/ lightning.htm [Accessed: January 10, 2019]

[12] Bolt of Lightning. A Memorial to Benjamin Franklin #1695, Direct Line Development [Internet]. Available from: http://freephillyphoto.com/ photos/1695/ [Accessed: January 10, 2019]

[13] Atmospheric Electricity [Internet]. 2019. Available from: https:// en.wikipedia.org/wiki/Atmospheric_ electricity [Accessed: January 10, 2019]

[14] Lightning Energy Storage System. US20140042987A1. 2014; Northern Lights Semiconductor Corp. Available from: https://patents.google. com/ patent/US20140042987A1/en [Accessed: January 10, 2019]

[15] Collecting and Storage Device of Lightning. CN106787229A. 2017; Anyang Teachers' College. Available from: https://patents.google.com/patent/ CN106787229A/en [Accessed: January 10, 2019]

[16] Xu H, He Y, Strobel KL, Gilmore CK, Kelley SP, Hennick CC, et al. Flight of an aeroplane with solid-state propulsion. Nature. 2018;**563**:532-535. Available from: https://www.nature. com/articles/s41586-018-0707-9 [Accessed: January 10, 2019] [17] Available from: http://news.mit. edu/2018/first-ionic-wind-plane-no-moving-parts-1121 [Accessed: January 10, 2019]

[18] Luo X, Wang J, Dooner M, Luo JCX, et al. Overview of current development in electrical energy storage technologies and the application potential in power system operation. Applied Energy. 2015;**137**:511-536. **Table 10**. Technical characteristics of electrical energy storage technologies

[19] Available from: http://www. nawatechnologies. com/en/nawa-technologies-revolutionizes- energy-storage-with-new-ultra-fast- carbon-battery/ [Accessed: January 10, 2019]

[20] Available from: http://www. rh.gatech.edu/features/12-emerging- technologies-may-help-power-future [Accessed: January 10, 2019]

[21] Available from: https://www. britannica.com/biography/Richard- Trevithick [Accessed: January 10, 2019]

[22] Michaelides EE. The Second Law of Thermodynamics as Applied to Energy Conversion Processes. Wiley Online Library; 1984. DOI: 10.1002/ er.4440080306

[23] EUR-Lex. Document 31980L0181. 1979. [Internet]. Available from: https:// eur-lex.europa.eu/legal-content/EN/ ALL/?uri=celex:31980L0181 [Accessed: January 10, 2019]

[24] Available from: https://en.wikipedia. org/wiki/Volt-ampere_reactive

[25] Introduction to Electronic Engineering. ISBN 978-87-7681-539-4. 2010; Valery Vodovozov & Ventus Publishing ApS

[26] Available from: http://www. ene.ttu.ee/elektria-jamid/oppeinfo/materjal/AAV0020/1Introduction1. pdf [Accessed: January 10, 2019]

[27] Qi J, Hahn A, Lu X, Wang J, Liu C-C. Cybersecurity for distributed energy resources and smart inverters. IET Cyber-Physical Systems: Theory & Applications. 2016;1(1):28. DOI: 10.1049/iet-cps.2016.0018. Available from: https://ieeexplore.ieee.org/stamp/stamp.jsp?arnumber=7805366

[28] Net Electricity Generation, EU-28 (% of total, based on GWh), Source: Eurostat (nrg_105a). 2016. Available from: https://ec.europa.eu/eurostat/ statistics-explained/index.php/ Electricity_production_consumption_ and_ market_overview [Accessed: January 10, 2019]

[29] Available from: http://appsso. eurostat.ec.europa. eu/nui/ submitViewTableAction.do [Accessed: January 10, 2019]

[30] Today in Energy, U.S. Energy Information Administration [Internet]. 2017. Available from: https://www.eia. gov/todayinenergy/detail.php?id=32912

[31] Available from: http://adsabs. harvard.edu/abs/1975AsAer..13...56B

[32] Available from: https://phys.org/ news/2018-07-national-ignition- facility-laser-energy.html

[33] The Nobel Prize in Physics 2014. NobelPrize.org. Nobel Media AB 2019. 2019. Available from: https://www. nobelprize.org/prizes/physics/2014/ summary/ [Accessed: January 10, 2019]

[34] Gales S, Tanaka KA, Balabanski DL, Negoita F, Stutman D, Tesileanu O, et al. The extreme light infrastructure- nuclear physics (ELI-NP) facility: New horizons in physics with 10 PW ultra-intense lasers and 20 MeV brilliant gamma beams. Reports on Progress in Physics. 2018;81(9):094301. DOI: 10.1088/1361-6633/aacfe8. Available from: https://www.ncbi.nlm.nih.gov/pubmed/29952755 [Accessed: January 10, 2019]

[35] Bahk S-W, Rousseau P, Planchon TA, Chvykov V, Kalintchenko G, Maksimchuk A, et al. Generation and characterization of the highest laser intensities (10^{22} W/cm^2). Optics Letters. 2004;29(24):2837. DOI: 10.1364/ ol.29.002837

[36] Available from: http://www. guinnessworldrecords.com/world- records/highest-intensity-focused-laser/ [Accessed: January 10, 2019]

[37] Sabu S, Srichitra S, Joby NE, Premlet B. Electric field characteristics during a thunderstorm: A review of characteristics of electric field prior to lightning strike. In: IEEE International Conference on Signal Processing, Informatics, Communication and Energy Systems (SPICES). 2017. DOI: 10.1109/spices.2017.8091357

[38] Rakov VA, Uman MA. Lightning: Physics and Effects. New York: Cambridge University Press; 2003. ISBN: 9780521583275

[39] Iarossi S, Poscolieri M, Rafanelli C, Franceschinis D, Rondini A, Maggi M, et al. The measure of atmospheric electric field. Lecture Notes in Electrical Engineering. 2011;91(3):175-179

[40] Available from: http://www.eli-np. ro/index.php

[41] Available from: https://www. nobelprize.org/uploads/2018/10/press- physics2018.pdf

[42] Yang L, Yu P, Wang B, Yao S, Liu Z. Review on cyber-physical systems. IEEE/CAA Journal of Automatica Sinica. 2017;4(1):27-40

[43] Kezunovic M, Annaswamy A, Dobson I, Grijalva S, Kirschen D, Mitra J, et al. Energy Cyber-Physical Systems: Research Challenges and Opportunities. 2016; National Science Foundation

[44] Lee EA, Cheng AMK. The past, present and future of cyber-physical systems: A focus on models. Sensors. 2015;15(3):4837-4869. DOI: 10.3390/ s150304837

[45] Available from: https://ptolemy. berkeley.edu/publications/papers/07/ PRET/ [Accessed: January 10, 2019]

[46] Available from: https://ptolemy. berkeley.edu/projects/summaries/11/ PTIDES.htm [Accessed: January 10, 2019]

[47] Jorge Juan Rodríguez Vázquez. Open CPS Platforms. 2015. DOI: 10.13140/RG.2.1.3161.4561. Available from: https://www.researchgate.net/ publication/280095935_Open_CPS_ Platforms [Accessed: January 10, 2019]

[48] Available from: http://roadmap2018. esfri.eu/ [Accessed: January 10, 2019]

[49] Available from: https://www. eurocps.org/ [Accessed: January 10, 2019]

[50] Available from: http://roadmap2018. esfri.eu/media/1066/esfri-roadmap-2018. pdf [Accessed: January 10, 2019]

[51] Available from: http://ec.europa. eu/research/in-frastructures/index.cfm [Accessed: January 10, 2019]

[52] Shepherd J. What Is the Digital Era? Source Title: Social and Economic Transformation in the Digital Era2004. DOI: 10.4018/978-1-59140-158-2.ch001.

Available from: https://www.igi- global.com/chapter/digital-era/29024 [Accessed: January 10, 2019]

[53] Robert F. Service, BIOELECTRONICS. The Cyborg Era Begins, Science. 07 Jun 2013;**340**(6137):1162-1165. DOI:10.1126/ science.340.6137.1162

4

Power Converter Topologies for Multiphase Drive Applications

Carlos A. Reusser

Abstract

The yet growing demand for higher demanding industrial applications and the global concern about harmful emissions in the atmosphere have increased the interest for new developments in electric machines and power converters. To meet these new requirements, multiphase machines have become a very attractive solution, offering potential advantages over three-phase classical solutions. Multiphase machine's power demand can be split over more than three phases, thus reducing the electric field stress on each winding (protecting the insulation system) and the requirements on maximum power ratings, for semiconductor devices. Moreover, only two degrees of freedom (i.e. two independently controllable currents) are required for independent flux and torque control. Due to the previous facts, the use of multiphase drives has become very attractive for applications and developments in areas such as electric ship propulsion, more-electric aircraft, electric and hybrid electric road vehicles, electric locomotive traction and in renewable electric energy generation. As a consequence of this multiphase drive tendency, the development of power converter topologies, capable of dealing with high power ratings and handling multiphase winding distributions, has encourage the development of new converter topologies, control strategies and mathematical tools, to face this new challenge.

Keywords: multiphase AC drive, neutral-point clamped, cascaded H-bridge, nine-switch converter, 11-switch converter, field-oriented control (FOC), direct torque control (DTC)

Introduction

Multiphase variable-speed drives, based on multiphase AC machines, are nowadays the most natural solution for high-demanding industrial, traction and power generation and distribution applications [1].

The types of multiphase machines for variable-speed applications are in principle the same as their three-phase counterparts. These are synchronous machines, which depending on the excitation can be subclassified into wound rotor, permanent magnet or reluctance machines and induction machines.

In a multiphase winding, the stator winding distribution becomes more concentrated, rather than distributed, as in the case of three-phase windings. This fact and the particularity of using quasi-sinusoidal, rather than sinusoidal voltages because of the inverting process in the power conversion stage, have several advantages that can be summarized as lower field harmonics content and better fault tolerance, because of extra degrees of freedom and less susceptibility to torque pulsations, due to an excitation field with less harmonic content [2].

As a consequence, the use of a multiphase winding configuration improves the MMF spatial distribution, by reducing its harmonic content and losses, due to flux leakage, and increasing the machine's efficiency. These facts have increased the interest for research and development for transportation applications, such as cargo ships, aircraft and road vehicles, thus also contributing to the reduction of green- house effect emissions.

Mathematical modeling of multiphase machines

All electrical machines can be found to be variations of a common set of funda- mental principles, which apply alike the number of phases of which the machine is constructed.

In this context, multiphase machines can be treated as belonging to an n- dimensional space, corresponding to the respective state variables. The machine model, on its original phase-variable form, can be transformed into $n/2$ two- dimensional subspaces, for a machine with an even number of phases. If the number of phases is odd, then the original n-dimensional subspace can be decomposed into a $(n - 1)/2$ two-dimensional subspaces and a single-dimensional quantity, which corresponds to a common mode subspace.

Each new two-dimensional subspace is orthogonal to each other, so there is no mutual coupling and they are represented in a stationary reference frame, respecting all other state variables. These new two-dimensional subspaces are could be denoted as u_k v_k, where k stands for the respective new orthogonal subspace.

Let us consider an arbitrary state variable λ, defined as $\lambda = [\lambda_1 \lambda_2 \dots \lambda_n]^T$, and let $T\langle\rangle$ be a linear operator, which transforms the n-dimensional space into $n/2$ two- dimensional subspaces, for n even, as in Eq. (1):

$$T\langle\rangle = \sqrt{\frac{2}{n}} \begin{bmatrix} 1 & \cos(\alpha) & \cos(2\alpha) & \dots & \cos(2\alpha) & \cos(\alpha) \\ 0 & \sin(\alpha) & \sin(2\alpha) & \dots & -\sin(2\alpha) & -\sin(\alpha) \\ 1 & \cos(2\alpha) & \cos(4\alpha) & \dots & \cos(4\alpha) & \cos(2\alpha) \\ 0 & \sin(2\alpha) & \sin(4\alpha) & \dots & -\sin(4\alpha) & -\sin(2\alpha) \\ \vdots & \vdots & \vdots & \vdots & \vdots & \vdots \\ 1 & \cos\left(\frac{n}{2}\right)\alpha & \cos 2\left(\frac{n}{2}\right)\alpha & \dots & \cos 2\left(\frac{n}{2}\right)\alpha & \cos\left(\frac{n}{2}\alpha\right) \\ 0 & \sin\left(\frac{n}{2}\right)\alpha & \sin 2\left(\frac{n}{2}\right)\alpha & \dots & -\sin 2\left(\frac{n}{2}\right)\alpha & -\sin\left(\frac{n}{2}\right)\alpha \end{bmatrix} \tag{1}$$

In the case of n odd, then the corresponding transformation is given as in Eq. (2):

$$
T\langle\rangle = \sqrt{\frac{2}{n}}
\begin{bmatrix}
1 & \cos(\alpha) & \cos(2\alpha) & \cdots & \cos(2\alpha) & \cos(\alpha) \\
0 & \sin(\alpha) & \sin(2\alpha) & \cdots & -\sin(2\alpha) & -\sin(\alpha) \\
1 & \cos(2\alpha) & \cos(4\alpha) & \cdots & \cos(4\alpha) & \cos(2\alpha) \\
0 & \sin(2\alpha) & \sin(4\alpha) & \cdots & -\sin(4\alpha) & -\sin(2\alpha) \\
\vdots & \vdots & \vdots & \vdots & \vdots & \vdots \\
1 & \cos\left(\frac{(n-1)}{2}\right)\alpha & \cos 2\left(\frac{(n-1)}{2}\right)\alpha & \cdots & \cos 2\left(\frac{(n-1)}{2}\right)\alpha & \cos 2\left(\frac{(n-1)}{2}\right)\alpha \\
0 & \sin\left(\frac{(n-1)}{2}\right)\alpha & \sin 2\left(\frac{(n-1)}{2}\right)\alpha & \cdots & -\sin 2\left(\frac{(n-1)}{2}\right)\alpha & -\sin 2\left(\frac{(n-1)}{2}\right)\alpha \\
\sqrt{\frac{1}{2}} & \sqrt{\frac{1}{2}} & \sqrt{\frac{1}{2}} & \cdots & \sqrt{\frac{1}{2}} & \sqrt{\frac{1}{2}}
\end{bmatrix}
\tag{2}
$$

The last row of T hi in Eq. (2) corresponds to the projection of the state variable onto the common mode subspace.

Using the above-presented mathematical decomposition, any machine can be represented by an equivalent two-axis idealized machine, called the Kron's primitive machine. The Kron's primitive machine corresponds to the representation of the n-dimensional space, in the $u_1 v_1$ or $\alpha\beta$ subspace, using the linear transformation $T\langle\rangle_{n\to\alpha\beta}$; this definition can be found as the generalized theory of electrical machines.

The mathematical model describing a generalized n-dimensional electrical machine is given in Eqs. (3) and (4), for dynamics, respectively:

$$
\boldsymbol{v_s} = [R_s]\,\boldsymbol{i_s} + \frac{d}{dt}\boldsymbol{\psi_s}
\tag{3}
$$

$$
\boldsymbol{v_r} = [R_r]\,\boldsymbol{i_r} + \frac{d}{dt}\boldsymbol{\psi_r}
\tag{4}
$$

$$
\boldsymbol{v_s} = \begin{bmatrix}v_{s1}v_{s2}\ldots v_{sn}\end{bmatrix}^T \qquad \boldsymbol{i_s} = \begin{bmatrix}i_{s1}i_{s2}\ldots i_{sn}\end{bmatrix}^T \qquad \boldsymbol{\psi_s} = \begin{bmatrix}\psi_{s1}\psi_{s2}\ldots\psi_{sn}\end{bmatrix}^T
\tag{5}
$$

$$
\boldsymbol{v_r} = \begin{bmatrix}v_{r1}v_{r2}\ldots v_{rn}\end{bmatrix}^T \qquad \boldsymbol{i_r} = \begin{bmatrix}i_{r1}i_{r2}\ldots i_{rn}\end{bmatrix}^T \qquad \boldsymbol{\psi_r} = \begin{bmatrix}\psi_{r1}\psi_{r2}\ldots\psi_{rn}\end{bmatrix}^T
\tag{6}
$$

Considering symmetrical windings for both stator and rotor, then $[R_s]$ and $[R_r]$ are diagonal $n \times n$ matrices, thus $[R_s] = diag\,(R_s)^{(n)}$ and $[R_r] = diag\,(R_r)^{(n)}$. The stator and rotor flux linkages can be found as in Eqs. (7) and (8):

$$
\psi_s = [L_s]\,i_s + [L_{sr}]\,i_r
\tag{7}
$$

$$
\psi_r = [L_r]\,i_r + [L_{rs}]\,i_s
\tag{8}
$$

where due to the machine's symmetry $[L_{rs}] = [L_{sr}]^T$; the corresponding stator and rotor inductance matrices are described as follows:

$$
[L_s] = \begin{bmatrix}
L_{s\,11} & L_{s\,12} & \cdots & L_{s\,1(n-1)} & L_{s\,1n} \\
\vdots & \vdots & \vdots & \vdots & \vdots \\
L_{s\,n1} & L_{s\,n2} & \cdots & L_{s\,n(n-1)} & L_{s\,nn}
\end{bmatrix}
\tag{9}
$$

$$[L_r] = \begin{bmatrix} L_{r11} & L_{r12} & \cdots & L_{r1(n-1)} & L_{r1n} \\ \vdots & \vdots & \vdots & \vdots & \vdots \\ L_{rn1} & L_{rn2} & \cdots & L_{rn(n-1)} & L_{rnn} \end{bmatrix} \tag{10}$$

It has to be noticed that for both stator and rotor windings, $L_{11} = L_{22} = \ldots = L_{nn}$ which correspond to the winding self-inductance L; for the corresponding stator and rotor self-inductances within each one, due to the machine's symmetry, it fulfills that $L_{jk} = L_{kj} = L_m \, \forall j, \, k \, j \neq k$. Then for the stator and rotor winding self-inductances, it can be stated that $L_{jj} = L_\sigma + L_m$ where L_σ corresponds to the stator or rotor winding leakage inductance. On the other hand, the stator-to-rotor mutual inductance matrix $[L_{sr}]$ can be found as in Eq. (11):

$$[L_{sr}] = M \begin{bmatrix} \cos(\theta_r) & \cos(\theta_r - (n-1)\alpha) & \cos(\theta_r - (n-2)\alpha) & \cdots & \cos(\theta_r - \alpha) \\ \cos(\theta_r - \alpha) & \cos(\theta_r) & \cos(\theta_r - (n-1)\alpha) & \cdots & \cos(\theta_r - 2\alpha) \\ \cos(\theta_r - 2\alpha) & \cos(\theta_r - \alpha) & \cos(\theta_r) & \cdots & \cos(\theta_r - 3\alpha) \\ \vdots & \vdots & \vdots & \cdots & \vdots \\ \cos(\theta_r - (n-1)\alpha) & \cos(\theta_r - (n-2)\alpha) & \cos(\theta_r - (n-3)\alpha) & \cdots & \cos(\theta_r) \end{bmatrix} \tag{11}$$

The resulting generalized machine representation is obtained by applying the linear transformation operator $T\langle \, \rangle_{n \to \alpha\beta}$ to Eqs. (3) and (4), resulting in the following dynamic representation, in the $\alpha\beta$ subspace, in Eqs. (12) and (13):

$$v_s^{(\alpha\beta)} = R_s i_s^{(\alpha\beta)} + \frac{d}{dt} \psi_s^{(\alpha\beta)} \tag{12}$$

$$v_r^{(\alpha\beta)} = R_s i_r^{(\alpha\beta)} + \frac{d}{dt} \psi_r^{(\alpha\beta)} - j\omega_r \psi_r^{(\alpha\beta)} \tag{13}$$

The corresponding stator and rotor flux linkages given in Eqs. (7) and (8) are transformed into the $\alpha\beta$ subspace as in Eqs. (14) and (15):

$$\psi_s^{(\alpha\beta)} = [L_s] \psi_s^{(\alpha\beta)} + [L_{sr}] i_r^{(\alpha\beta)} \tag{14}$$

$$\psi_r^{(\alpha\beta)} = [L_r] i_r^{(\alpha\beta)} + [L_{rs}] i_s^{(\alpha\beta)} \tag{15}$$

The presented methodology considers a n-phase winding with uniform distribution that is $\alpha = \frac{2\pi}{n}$; in the case of machine with multiple groups of three-phase windings, α is the shifting angle between each group of three-phase windings. It has to be noted that for each three-phase winding transformation, an additional row has to be included in Eq. (2).

The developed electromechanical torque can be expressed in terms of the electromechanical energy conversion as in Eqs. (16) and (17):

$$T_e = \frac{\partial}{\partial \theta_r} W_{fld} \left(i_s^{(\alpha\beta)}, \psi_s^{(\alpha\beta)}, \theta_r \right) \tag{16}$$

$$T_e = \frac{1}{2} \left[i^{(\alpha\beta)} \right] \frac{\partial}{\partial \theta_r} [L] \left[i^{(\alpha\beta)} \right]^T \tag{17}$$

$$\left[i^{(\alpha\beta)} \right] = \left[i_s^{(\alpha\beta)} \quad i_r^{(\alpha\beta)} \right]^T \tag{18}$$

$$[L] = \begin{bmatrix} [L_s] & [L_{sr}] \\ [L_{sr}]^T & [L_r] \end{bmatrix} \tag{19}$$

Following the electromechanical torque can be expressed as in Eq. (20):

$$T_e = \frac{1}{2} i_s^{(\alpha\beta)} \frac{\partial}{\partial \theta_r} [L_{sr}] i_r^{(\alpha\beta)^T} \tag{20}$$

The generalized machine representation given by Eqs. (12), (13) and (20) can be expressed in an arbitrary orthogonal synchronous reference frame dq, which rotates at synchronous speed ω_k; thus the corresponding rotation into the dq state variables is given by Eq. (21):

$$\theta_k = \int \omega_k \, dt \tag{21}$$

Rotation into the dq subspace is given by the unitary rotation matrix U defined in Eq. (22):

$$U = \begin{bmatrix} \cos \theta_k & \sin \theta_k \\ -\sin \theta_k & \cos \theta_k \end{bmatrix} \tag{22}$$

The generalized machine dynamic representation in the dq reference frame, as consequence of the rotation of Eqs. (12), (13) and (20), is given in Eqs. (23)–(25), representing the Kron's primitive machine model in a generalized synchronous reference frame:

$$v_s^{(dq)} = R_s i_s^{(dq)} + \frac{d}{dt} \psi_s^{(dq)} + j \omega_k \psi_s^{(dq)} \tag{23}$$

$$v_r^{(dq)} = R_s i_r^{(dq)} + \frac{d}{dt} \psi_r^{(dq)} + j (\omega_k - \omega_r) \psi_r^{(dq)} \tag{24}$$

$$T_e = p \frac{L_m}{L_r} \left(\psi_r^d i_s^q - \psi_r^q i_s^d \right) \tag{25}$$

Equations (23) and (24) can be written in their matrix form as in Eq. (26), becoming a generalized impedance model, as described in Eq. (27):

$$\begin{bmatrix} v_s^d \\ v_s^q \\ v_r^d \\ v_r^q \end{bmatrix} = \begin{bmatrix} R_s + L_s \dfrac{d}{dt} & -\omega_k L_s & L_m \dfrac{d}{dt} & -\omega_k L_m \\ \omega_k L_s & R_s + L_s \dfrac{d}{dt} & \omega_k L_m & L_m \dfrac{d}{dt} \\ L_m \dfrac{d}{dt} & -(\omega_k - \omega_r) L_m & R_r + L_r \dfrac{d}{dt} & -(\omega_k - \omega_r) L_r \\ (\omega_k - \omega_r) L_m & L_m \dfrac{d}{dt} & (\omega_k - \omega_r) L_r & R_r + L_r \dfrac{d}{dt} \end{bmatrix} \begin{bmatrix} i_s^d \\ i_s^q \\ i_r^d \\ i_r^q \end{bmatrix} \tag{26}$$

$$\left[v^{(dq)} \right] = [z] \left[i^{(dq)} \right] \tag{27}$$

Multiphase synchronous machines

The use of multiphase synchronous machines has been focused on its application for medium and high-power generation systems, being their primary use, on wind energy conversion systems (WECS). The previous statement is based on the fact that most of wind energy conversion systems operate in the low-voltage range, principally due to restrictions of winding insulation. This fact has stimulated the development of multiphase generator topologies, thus gaining increasing interest in the research for new converter topologies [3, 4]. Some of their main advantages are the following:

- The total power can be divided into lower power converters.

- Due to that each converter is insulated from each other, there is no circulating current between converters, which leads to no power derating for the converters.

- The increase of the number of phases in the generator voltages is phase-shifted so that low-order harmonics are reduced and consequently smaller filters can be used.

- Reduced torque pulsations.

- Fault-tolerant redundancy under winding fault conditions.

- Additional degrees of freedom which can be used to improve the machine performance.

Arrangement of multiple three-phase windings has become a very popular construction technique, for multiphase machines. In this field, the most common configuration is the six-phase machine, based on two independent three-phase windings, which are spatially shifted in 30 degrees, as shown in **Figure 1**.

The use of multiple three-phase windings has the advantage of guaranteeing full decoupling under faulty conditions, thus preventing the circulation of common mode currents and pulsating torque.

The linear transformation operator T hi is defined as in Eq. (28), with $\varphi = \pi/6$:

$$T\langle\rangle = \frac{1}{\sqrt{3}} \begin{bmatrix} 1 & \cos 4\varphi & \cos 8\varphi & \cos \varphi & \cos 5\varphi & \cos 9\varphi \\ 0 & \sin 4\varphi & \sin 8\varphi & \sin \varphi & \sin 5\varphi & \sin 9\varphi \\ 1 & \cos 8\varphi & \cos 4\varphi & \cos 5\varphi & \cos \varphi & \cos 9\varphi \\ 0 & \sin 8\varphi & \sin 4\varphi & \sin 5\varphi & \sin \varphi & \sin 9\varphi \\ 1 & 1 & 1 & 0 & 0 & 0 \\ 0 & 0 & 0 & 1 & 1 & 1 \end{bmatrix} \tag{28}$$

The equivalent model in the dq synchronous subspace can be derived from Eq. (26); in this case it has to be noted that the synchronous reference frame dq coincides with the rotor natural reference frame, thus $\omega_k = \omega_r$, becoming Eq. (29) and (30) for the stator:

$$v_s^d = R_s i_s^d + \frac{d}{dt} \psi_s^d - \omega_k \psi_s^q \tag{29}$$

$$v_s^q = R_s i_s^q + \frac{d}{dt} \psi_s^q + \omega_k \psi_s^d \tag{30}$$

Equations (31) and (32) correspond to the damping winding effect:

$$0 = R_r i_r^d + \frac{d}{dt} \psi_r^d \tag{31}$$

$$0 = R_r i_r^q + \frac{d}{dt} \psi_r^q \tag{32}$$

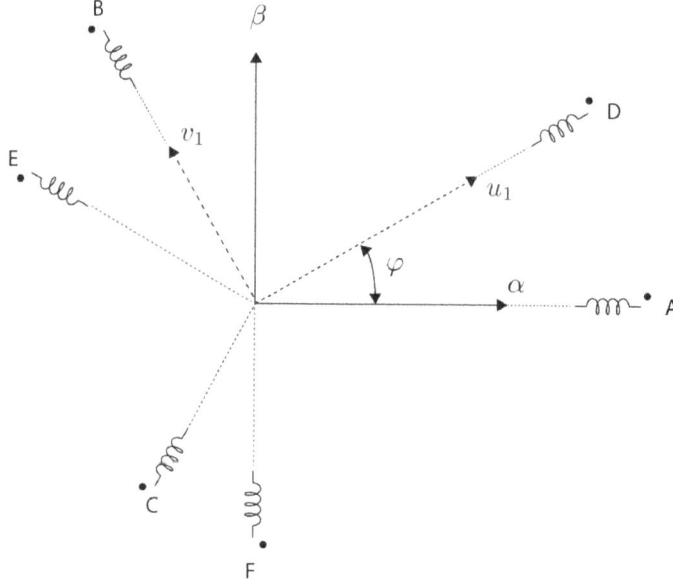

Figure 1. *Dual three-phase stator winding configuration.*

The corresponding flux linkages, considering the general case of an anisotropic machine, are given in the following equations:

$$\psi_s^d = L_{sd} i_s^d + L_{md} i_r^d + \psi_m \tag{33}$$

$$\psi_s^q = L_{sq} i_s^q + L_{mq} i_r^q \tag{34}$$

$$\psi_r^d = L_{rd} i_r^d + L_{md} i_s^d + \psi_m \tag{35}$$

$$\psi_r^q = L_{rq} i_r^q + L_{mq} i_s^q \tag{36}$$

and the electromechanical torque is given in Eq. (37):

$$T_e = p \left[\psi_m i_s^q + \left(L_{md} i_r^d i_s^q - L_{mq} i_r^q i_s^d \right) \right] + p \left[L_{md} - L_{mq} \right] i_s^q i_s^d \tag{37}$$

Multiphase induction machines

Modern industrial high-demanding processes are commonly based on induction machines. They are very attractive for these kind of applications because of their simplicity and capability to work under extreme torque demanding conditions [5].

Within the previous context, multiphase induction machines have become very popular for applications where high redundancy and power density are required. In particular, the use of multiple three-phase windings (six and nine phases) in naval propulsion systems has aroused much interest and encouraged the research of new multiphase converter topologies and control schemes [6, 7].

Based on the Kron's primitive machine model developed in Eq. (26), the dynamical model of the multiphase induction machine is given in Eqs. (38)–(41):

$$v_s^d = R_s i_s^d + \frac{d}{dt} \psi_s^d - \omega_k \psi_s^q \tag{38}$$

$$v_s^q = R_s i_s^q + \frac{d}{dt} \psi_s^q + \omega_k \psi_s^d \tag{39}$$

$$0 = R_r i_r^d + \frac{d}{dt} \psi_r^d \tag{40}$$

$$0 = R_r i_r^q + \frac{d}{dt} \psi_r^q \tag{41}$$

Due to the isotropy of the induction machine and the absence of an MMF source in the rotor (permanent magnets or field winding), the corresponding flux linkages are given in Eqs. (42)–(45):

$$\psi_s^d = L_s i_s^d + L_m i_r^d \tag{42}$$

$$\psi_s^q = L_s i_s^q + L_m i_r^q \tag{43}$$

$$\psi_r^d = L_r i_r^d + L_m i_s^d \tag{44}$$

$$\psi_r^q = L_r i_r^q + L_m i_s^q \tag{45}$$

The electromechanical torque expression can be derived from Eq. (37) by considering $\psi_m = 0$, $L_{md} = L_{mq} = L_m$, becoming Eq. (46):

$$T_e = p L_m \left[i_r^d i_s^q - i_r^q i_s^d \right] \tag{46}$$

Multiphase power converters

As stated previously, multiphase machines have many advantages over traditional three-phase-based machine drives, by reducing the impact of low-frequency torque pulsations and the dc-link current harmonic content. They also, due to the nature of their winding configuration, improve the system reliability, by introducing redundant operation conditions. As a consequence the use of multiphase converter topologies with multiphase drive arrangements has been proved as a viable approach for its application in high-demanding industrial applications.

In this field, the development of power converters capable to deal with the multiphase machine structure has capture much attention in the recent years; thus several topologies have been introduced in the last decades. This topologies can consist of arrangements of conventional two-level three-phase voltage source converters (2LVSC), multilevel converters (MLVSC) or in more specialized and dedicated topologies such as multiphase matrix converters.

Classical topologies

Classical topologies for multiphase converters are commonly based in arrangements of parallel-connected fundamental cells (multiple legs), or in multiple channel configuration, of voltage source converter topologies.

Commonly used topologies for multiphase applications are H-bridge converter (HBC), neutral-point clamped converter (NPC) and the cascaded H-bridge (CHB) topology, which are shown in **Figure 2**.

Classical multiphase VSC topologies consist of an arrangement of n individual half-cells connected in parallel to a single dc-link, as shown in **Figure 3**.

Modulation of each individual half-cell circuit is implemented in such a way, to obtain n voltage output signals shifted by $2\pi/n$. Generation of the required voltage signals is supported by classical carrier-based strategies, such as sinusoidal pulse width modulation (SPWM) and space vector pulse width modulation (SVPWM).

SPWM methods can be implemented for two-level or multilevel half-cells. In the first case, only one high-frequency carrier is required for the respective switching signals per phase S_x, as given in Eqs. (46) and (48):

$$S_x = \begin{cases} 1 & |u_{xN}^*| \ge |u_c| \\ 0 & |u_{xN}^*| < |u_c| \end{cases} \tag{47}$$

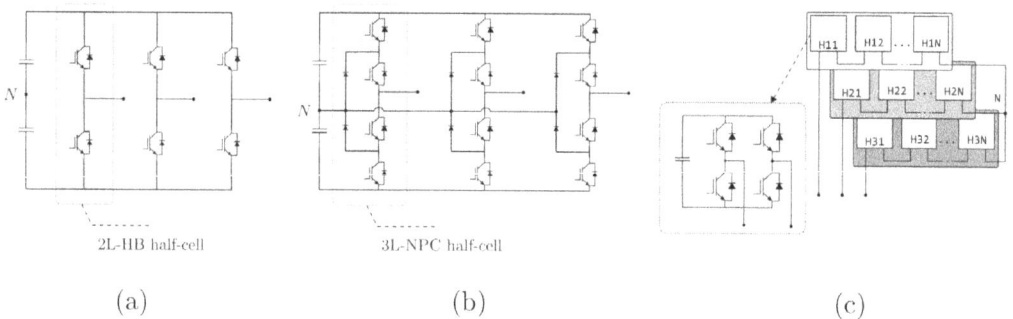

2L-HB half-cell	3L-NPC half-cell	
(a)	(b)	(c)

Figure 2. *Classical VSC topologies: (a) two-level bridge, (b) neutral-point clamped and (c) cascaded H-bridge.*

$$u_{xN}^* = v_{xN}^* - \frac{1}{2}\left\{ min\left(v_{aN}^*, v_{bN}^*, \ldots, v_{nN}^*\right) + max\left(v_{aN}^*, v_{bN}^*, \ldots, v_{nN}^*\right) \right\} \tag{47}$$

where x stands for the corresponding phase x = [1; … ;n], u_{xN}^* is the reference to be synthesized for the corresponding phase, and u_c stands for the carrier wave.

For multilevel parallel arrangements (NPC or CHB), the use of multiple carries, as an extension of the two-level PWM methods, has been proven as a suitable solution. Level-shifted PWM

(LSPWM) has become a very popular modulation technique, because it fits for any multilevel converter topology and ensures low harmonic distortion. The corresponding switching states are given in Eq. (49):

$$S_{xm} = \begin{cases} 1 & |u^*_{xN}| \geq |u_{cm}| \\ 0 & |u^*_{xN}| < |u_{cm}| \end{cases} \tag{49}$$

where x stands for the corresponding phase $x = [\, 1; \ldots ; n\,]$, m to the corresponding level and, u_{cm} is the carrier for the matched m level.

The main advantages of this topology are its simplicity by using half-bridges for each leg (phase), and the requirement of a single dc-link the feed the n-phase inverting stage. The use of a parallel arrangement of fundamental half-cells generates $(n - 1)$ dimensional spaces, which can be decomposed into $(n - 1)/2$ subspaces, ensuring redundant switching states.

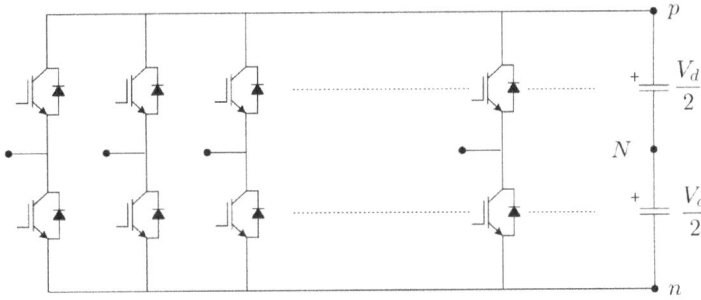

Figure 3. *Multiphase multicell topology.*

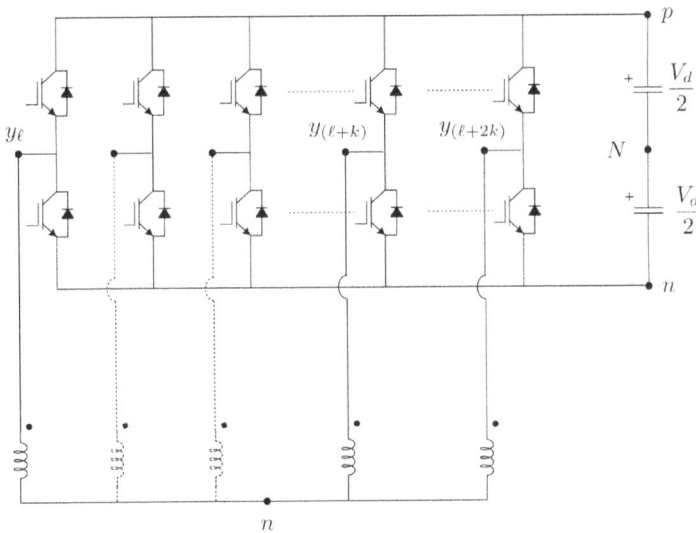

Figure 4. *Common connected load arrangement.*

Multicell topology also enables the possibility to have common or split connected loads. In the first case, each phase in the load side is connected to a common neutral point N as presented in **Figure 4**; the main drawback with this configuration is the circulation of zero sequence currents, since the neutral points N - n are not insulated, thus establishing a common mode voltage v_{Nn} 6 = 0 in the dq synchronous reference frame.

Split connected load arrangement is possible, if $n = 3 k \mid k = [\ 1;\ 2;\ \ldots\ ]$, so the n phase system can be divided into k three-phase insulated independent subsystems, as shown in **Figure 5**, with symmetric or asymmetric electrical shifting δ. It has to be noted that independently on the symmetric or asymmetric configuration, the electrical shifting between each phase on a single three-phase group is still $2\pi/3$, constituting each individual symmetrical three-phase system.

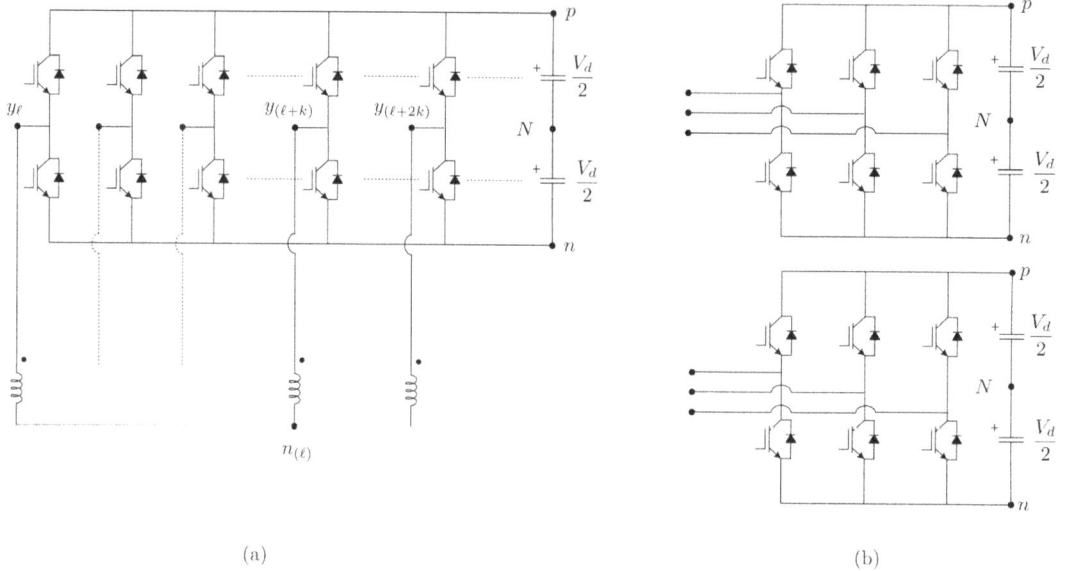

(a) (b)

Figure 5. *Split connected load arrangements. (a) Single-channel topology and (b) multichannel topology.*

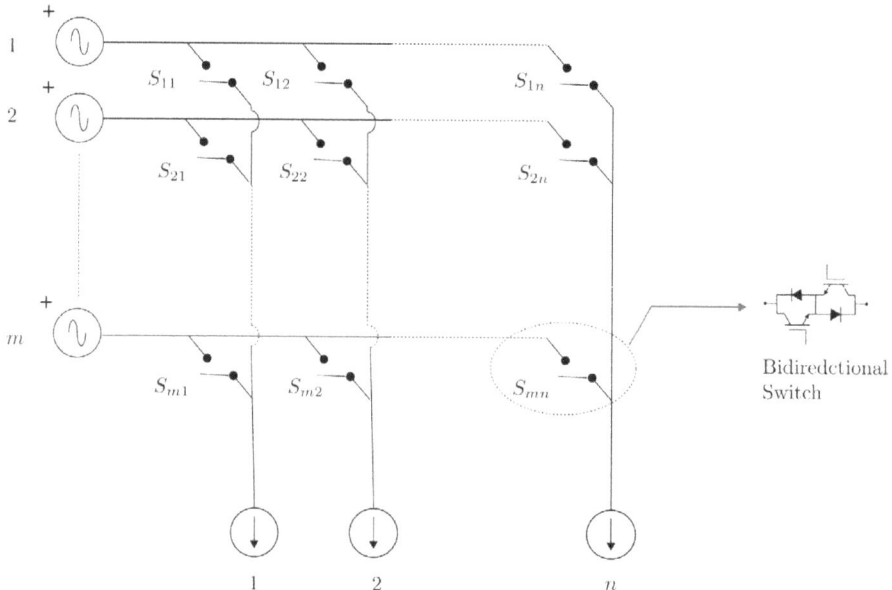

Figure 6. *mxn matrix converter equivalent model.*

Figure 6a shows a split-phase single-channel arrangement where the corresponding phase for each half-cell is derived as $y_\ell, y_{\ell+k}, y_{+2k}$ with $\ell = [\ 1;\ \ldots\ ;\ k\ ]$.

A multichannel arrangement is shown in **Figure 6b**, which introduces some advantages, like full load dynamic decoupling and lower dc-link power rating. On the other hand, additional dc-link capacitors are required.

For both previously described topologies, SPWM is achieved by implementing the modulation as given in Eqs. (47) and (48) considering the asymmetry (if required) φ for each reference as $u_{xN}^{*}(\varphi)$.

In the case of space vector modulation for multiphase converters, it is necessary to extend the space vector representation into its corresponding subspaces. So as presented previously in this chapter, the output voltage space vector can be decomposed into $n/2$ orthogonal subspaces for n even and into $(n - 1)/2$ orthogonal subspaces and single-dimensional quantity (common mode component) for n odd. Under this formulation, the reference voltage space vector to be synthesized can be expressed as in Eqs. (50) and (51) for n even and odd, respectively:

$$u_x^* = \left[u^{(1)} u^{(2)} \ldots , u^{(n)} \right]^T \tag{50}$$

$$u_x^* = \left[u^{(1)} u^{(2)} \ldots , u^{(n-1)} \right]^T \tag{51}$$

For each orthogonal subspace, there exits two active vectors, $\bar{v}_{uk}$ and $\bar{v}_{vk}$, corresponding to k orthogonal subspace. Each active vector is applied for a duty-cycle δ to synthesize the corresponding reference space vector, as in Eqs. (52) and (53):

$$u_x^* = \delta_{u1}\bar{v}^{(u1)} + \delta_{v1}\bar{v}^{(v2)} + \delta_{u2}\bar{v}^{(u2)} + \delta_{v2}\bar{v}^{(v2)} + \ldots + \delta_{u(n/2)}\bar{v}^{(u(n/2))} + \delta_{v(n/2)}\bar{v}^{(v(n/2))} \tag{52}$$

$$u_x^* = \sum_{j=1}^{n/2} \delta_{uj}\bar{v}^{(uj)} + \sum_{j=1}^{n/2} \delta_{vj}\bar{v}^{(vj)} \tag{53}$$

In the case of n odd, an additional degree of freedom is introduced by the common mode voltage v_{cm} which represents a single quantity and not a space vector as in Eq. (54):

$$u_x^* = \sum_{j=1}^{(n-1)/2} \delta_{uj}\bar{v}^{(uj)} + \sum_{j=1}^{(n-1)/2} \delta_{vj}\bar{v}^{(vj)} + \delta_{cm}v_{cm} \tag{54}$$

Eqs (53) and (54) can be interpreted as the partitioning of the $\alpha\beta$ stationary reference frame, into $n!$ or $(n - 1)!$ adjacent sectors, and an additional common mode component, orthogonal to the $\alpha\beta$ subspace. Sector identification can then be achieved by implementing Eq. (55) where S corresponds to the given sector within the $\alpha\beta$ subspace:

$$S = \left(\frac{\theta}{M!} \right) + 1 \quad M = \begin{cases} n & \text{even number of subspaces.} \\ (n-1) & \text{odd number of subspaces.} \end{cases} \tag{55}$$

Matrix converter

The matrix converter is a direct AC-AC converter, which uses an arrangement of bidirectional switches, to connect each input phase, with a single corresponding output phase, thus generating an arrangement of mxn power switches, where m is the number of input phases and n the number of output phases. In **Figure 6**, the equivalent model of a mxn matrix converter is shown.

Due to the absence of a dc-link stage, the output voltages should be synthesized by selecting segments of the input voltages, by generating the adequate switching states. However, some switching state restrictions have to be taken into account, because of the particular topology of the matrix converter. Let's consider a generalized switching state S_{jk}, such that

$$S_{jk} = \begin{cases} 1 & \text{switch } S_{jk} \text{ is in on state.} \\ 0 & \text{switch } S_{jk} \text{ is in off state.} \end{cases} \quad \forall j = 1, 2, \ldots, m \quad k = 1, 2, \ldots, n \tag{56}$$

For every instant t, the switching state S_{jk} must comply with both conditions stated in Eqs (57) and (58), thus meaning that for all t only one input phase is connected to one output phase, avoiding short-circuit condition, and also all output phases are connected to at least one input phase, ensuring no open-circuit condition. This last condition is intended to protect the bidirectional switches that cannot handle reverse current flow, due to the inductive energy discharge:

$$\sum_{j=1}^{n} S_{jk} = 1 \tag{57}$$

$$S_{jk} = 1 \quad \forall k = 1, 2, \ldots, n \tag{58}$$

Sinusoidal PWM scheme is implemented via the *Venturini* method [8], which is based on the solution of the relational input-output equations of the matrix converter, given in Eqs. (59)–(61):

$$[\mathbf{V_o}] = [M][\mathbf{V_i}] \quad M = \begin{bmatrix} \delta_{11} & \delta_{12} & \ldots & \delta_{1m} \\ \delta_{21} & \delta_{22} & \ldots & \delta_{2m} \\ \vdots & \vdots & \ldots & \vdots \\ \delta_{n1} & \delta_{n2} & \ldots & \delta_{nm} \end{bmatrix} \tag{59}$$

$$[\mathbf{I_i}] = [M]^T [\mathbf{I_o}] \tag{60}$$

$$P_i = P_o \tag{61}$$

where $[\mathbf{V_o}]$ and $[\mathbf{V_i}]$ stand for the output and input voltage space vectors, respectively, $[\mathbf{I_o}]$ and $[\mathbf{I_i}]$ for the output and input current space vectors, $[M]$ is the low-frequency matrix or modulation index transfer matrix, and P_o and P_i corre- spond to the output and input active power.

On the other hand, space vector PWM formulation has no difference, as it is implemented in VSC. However, its complexity lies in the fact in that the absence of a nearly constant dc-link voltage, from which the reference voltage space vectors are synthesized. So both space vectors are to be composed using the input and output, voltage and current space vectors simultaneously. Moreover, the total number of possible switching states is $2^{(nm)}$, from which the forbidden conditions have to be considered. Space vector implementation is largely explained in literature [9].

The matrix converter has several advantages over multiphase voltage source converters, for multiphase drive applications. It is capable to synthesize nearly sinusoidal output voltage and currents, with low-order harmonics, thus improving the MMF distribution in machine air gap, eliminating torque ripple and preventing mechanical stresses on the output shaft. It provides bidirectional energy flow and provides full power factor control. Moreover, due to the absence of dc-link, it presents more power density, because of the lack of large capacitors, becoming an alternative for integrated drive converter configurations [10, 11].

However the complexity in the implementation of SPWM and SVPWM schemes, and the complex commutation strategy for the bidirectional switches, makes the matrix converter less attractive than the voltage source-based multiphase solutions.

There exist several variations of the direct matrix converter topology presented previously, such as the indirect and sparse matrix converter topologies, which are extensively described in literature [12]. These topological variations relay on the same basis but differ in the number and type of power switches used.

Nine-switch converter

The topology is derived from two three-phase voltage source converters that share a positive and a negative busbar, respectively, as shown in **Figure 7**.

The nine-switch converter has the ability to operate in back-to-back mode, as rectifier (A stage input) and inverter (B stage output), as two channel rectifier (A and B stages input) and as two stage inverter (A and B stages output), enabling the converter to handle a six-phase systems with just one channel, in spite of commonly 12-switch back-to-back topologies.

However, this topological advantage introduces some drawbacks to the nine- switch topology, because of the fact that S_x, S_y and S_z switches are shared by both converter stages A and B. Hence 3 switches are eliminated of the 12 needed for multiphase operation as in multichannel topology, some forbidden switching states are introduced, thus remaining 27 allowed states, and also the maximum output voltage gain is limited [13]. In **Figure 8**, the allowed switching states per leg are shown.

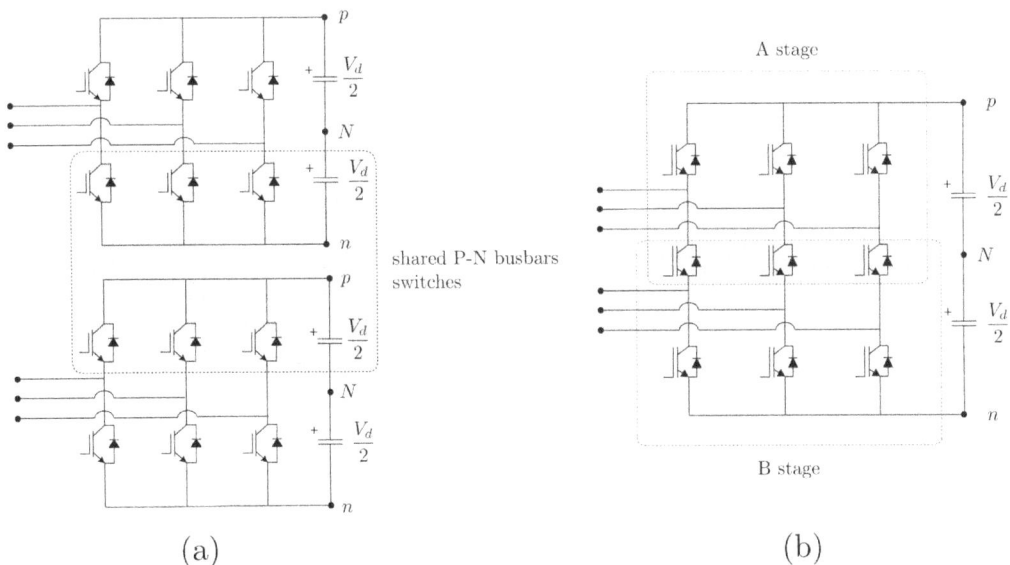

(a)   (b)

Figure 7. (a) Classical arrangement with 12 switches. (b) Nine-switch converter topology.

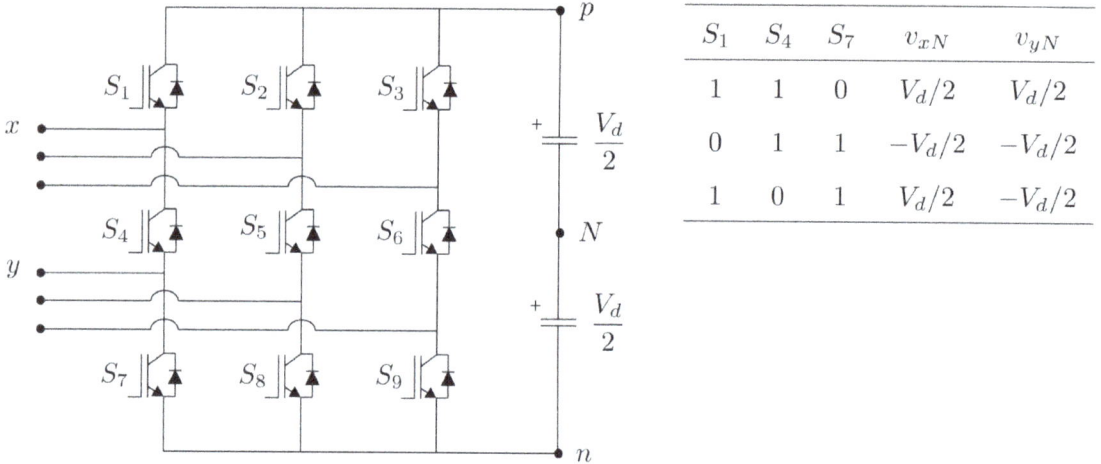

Figure 8. *Nine-switch converter allowed switching states.*

S_1	S_4	S_7	v_{xN}	v_{yN}
1	1	0	$V_d/2$	$V_d/2$
0	1	1	$-V_d/2$	$-V_d/2$
1	0	1	$V_d/2$	$-V_d/2$

Carrier-based PWM modulation schemes such as sinusoidal PWM (SPWM), space vector modulation (SVM) and min-max (third harmonic injection) PWM [14] can be applied to the nine-switch converter, using two independent SPWM modulation schemes, one for each converter stage, as in the 12-switch back-to-back converter. However, because of the restrictions introduced to the modulation, pattern by the middle switches S_4, S_5 and S_6, its switching pattern can be obtained as in Eq. (62):

$$S_{k+3} = \overline{S_k \cdot S_{k+6}} \quad k = 1, \ldots, 3 \tag{62}$$

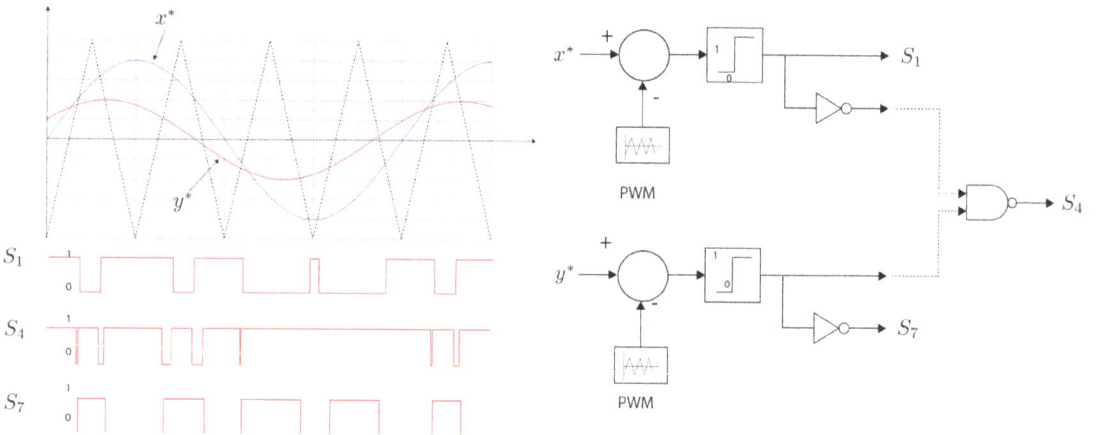

Figure 9. *CFM modulation.*

The implementation of SPWM considers the use of two voltage reference space vectors $x*$ and $y*$, for each three-phase winding group. The NSC can operate either in constant (CFM) or variable frequency (VFM) modes, depending on each space vector angular frequency. However, due to the nature of the particular application in multiphase drives, the NSC is to be operated in the constant frequency mode, as presented in **Figure 9**.

For the constant frequency mode operation (CFM), both reference space vectors $x*$ and $y*$ have the same angular frequency ω_k, preserving its spatial shifting δ, so they are to be found at least in contiguous sectors of the voltage hexagon, as shown in **Figure 10a**, being the maximum voltage gain g given in Eq. (63):

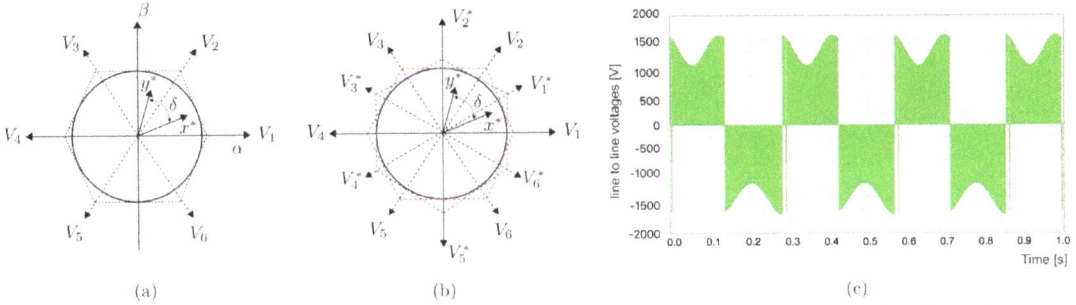

Figure 10. *Space vector decomposition. (a) Single carrier, (b) shifted multi-carrier and (c) line-line voltage.*

$$g = \frac{1}{\sqrt{3}\left(\frac{\delta}{2} + \frac{\pi}{3}\right)} \tag{63}$$

By introducing a phase shifting of π/6 between both PWM carriers, a second group of active vectors is introduced as shown in **Figure 10b**, which results in a voltage gain as in Eq. (64):

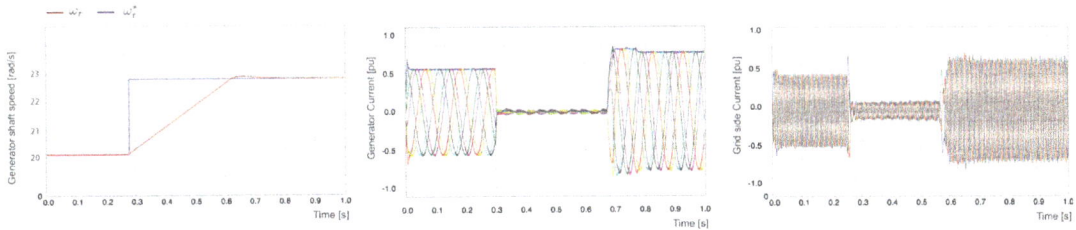

Figure 11. *Simulation results for a back-to-back NSC for wind energy conversion.*

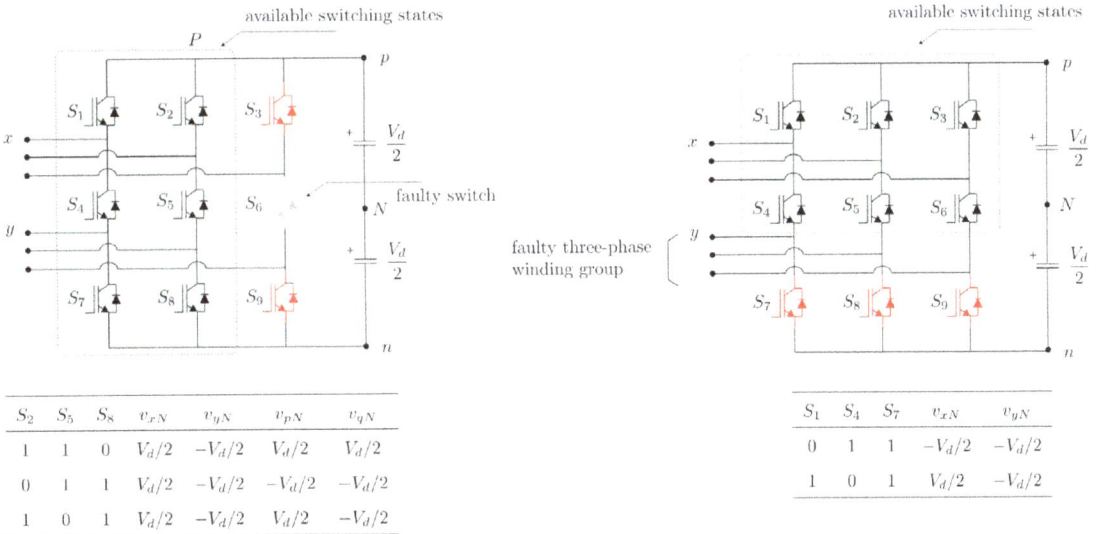

S_2	S_5	S_8	v_{xN}	v_{yN}	v_{pN}	v_{qN}
1	1	0	$V_d/2$	$-V_d/2$	$V_d/2$	$V_d/2$
0	1	1	$V_d/2$	$-V_d/2$	$-V_d/2$	$-V_d/2$
1	0	1	$V_d/2$	$-V_d/2$	$V_d/2$	$-V_d/2$

S_1	S_4	S_7	v_{xN}	v_{yN}
0	1	1	$-V_d/2$	$-V_d/2$
1	0	1	$V_d/2$	$-V_d/2$

Figure 12. *Nine-switch converter reconfiguration under faulty operation.*

Recently, the NSC has been gaining much attention in various applications like isolated wind-hydro hybrid power system, power quality enhancement and hybrid electric vehicles because of its ability to interconnect multiphase power systems in common or split configurations and also independent three-phase-based systems, in constant or variable frequency modes. **Figure 11** shows simulation results for a wind energy conversion system (WECS) based on a nine-switch back-to-back converter topology [15].

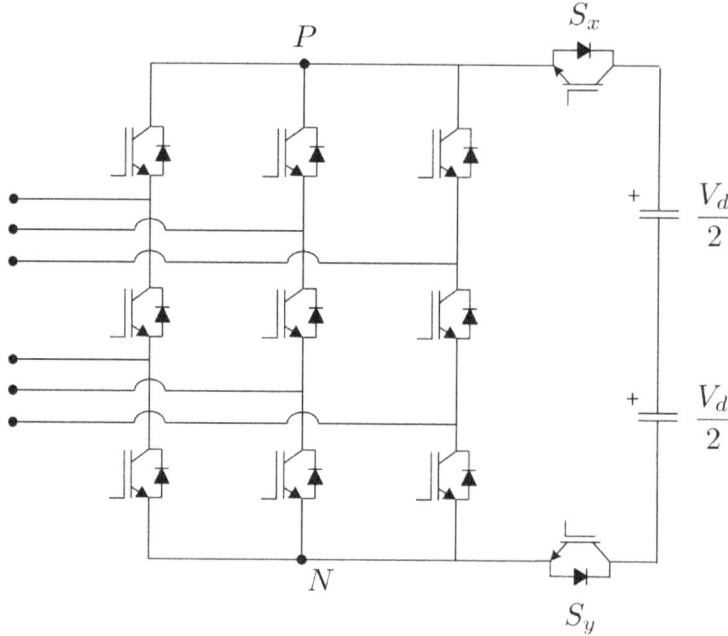

Figure 13. *11-switch converter.*

$$g = \frac{1}{\sqrt{3}\left(\delta + \frac{\pi}{6}\right)} \tag{64}$$

Another important feature of the NSC is the capability to rearrange its switching states under single- or multiple-switch faulty conditions [16], as shown in **Figure 12**.

11-switch converter

The 11-switch converter, presented in [17], consists of a modified topology of the 9-switch converter topology, previously discussed. As shown in **Figure 13**, this topology introduces two additional switches S_x, S_y, whose main propose is to mitigate the common mode voltage during the zero switching states.

Control of multiphase electric drives

As discussed previously in this chapter, the electromechanical torque developed by the multiphase machine depends only on the state variables in $\alpha\beta$ subspace, thus meaning that the other subspaces do not contribute to the energy conversion process but only losses. This fact makes possible the implementation of oriented control schemes, such as field-oriented control (FOC), direct torque control (DTC) and model-based predictive control (MBPC), in a synchronous reference frame dq, by rotating the state variables in $\alpha\beta$ subspace.

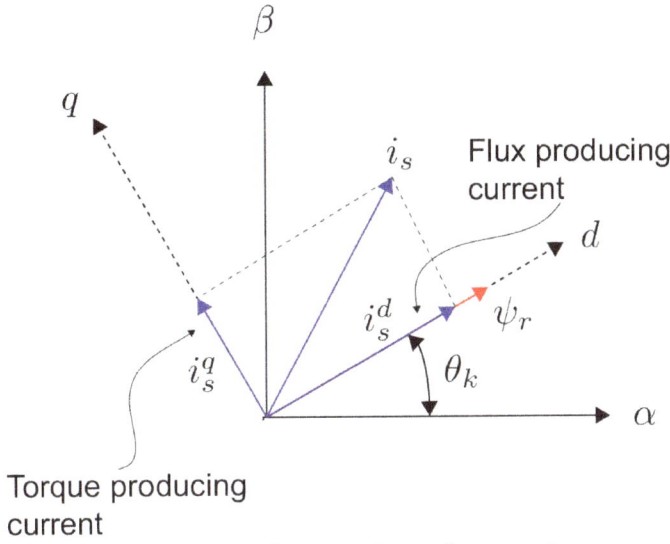

Figure 14. *dq synchronous reference frame rotation.*

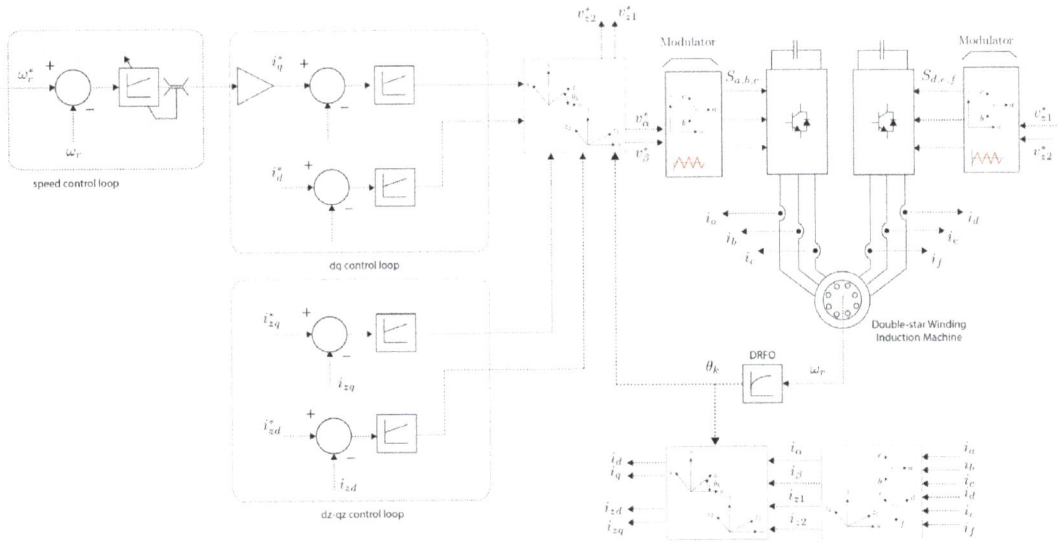

Figure 15. *Field-oriented control scheme of a dual three-phase winding induction machine.*

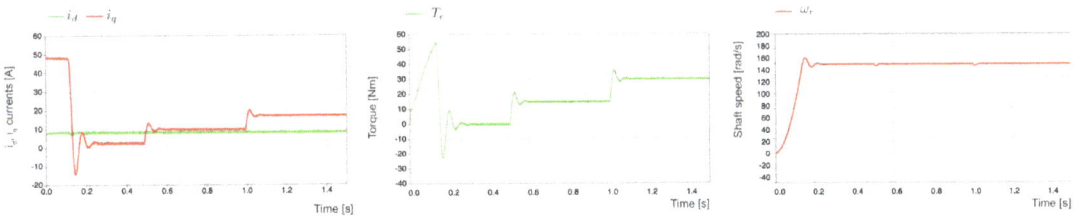

Figure 16. *Dual three-phase winding induction machine FOC scheme simulation results.*

An additional control loop in the *dq* reference frame can be implemented, to compensate the losses in the remaining subspaces. Control goals for the multiphase drive can be summarized as maximum torque per Ampere operation, control of nominal flux and control of rotor speed/torque.

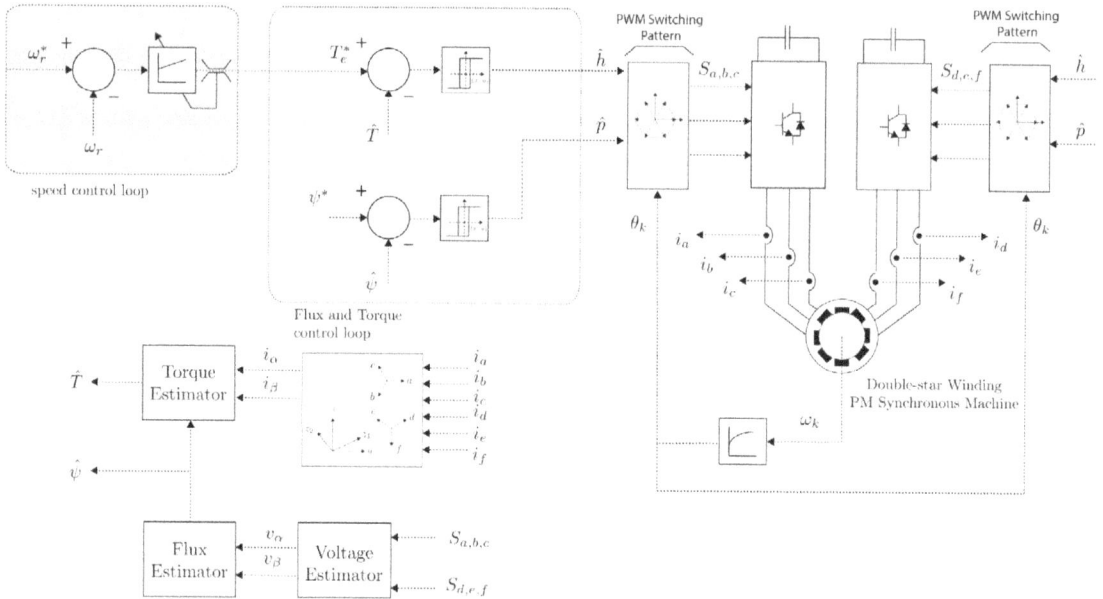

Figure 17. *Direct torque control scheme of a dual three-phase winding PM synchronous machine.*

Field-oriented control

The characteristics of field-oriented control (FOC) have made this control strategy the most widely used for high-demanding industrial applications. Field- oriented control is based in the decoupling of the current space vector, into a flux- producing component and a torque-producing component. This is achieved by rotating the current space vector from $\alpha\beta$ subspace, into the synchronous rotating reference frame dq, oriented with respect to the rotor flux linkage space vector, as shown in **Figure 14**.

In this way the magnetizing flux can be controlled, so that the machine operates with nominal flux under any condition, and also torque can be controlled only by i_s^q, because $\psi_r^q = 0$, due to the orientation of ψ_r which only exists in the direct axis orientation. Implementation of the FOC scheme for a multiphase AC drive is shown in **Figure 15**.

Figure 16 shows simulation results for a dual three-phase induction machine drive using FOC scheme, under load impact.

The main drawbacks of FOC are the requirement in the estimation of the rotor flux linkage space vector and the rotation of the state space variables into this synchronous reference frame, becoming a very complex process. Also, as the dynamic response, this control strategy is limited by maximum bandwidth achievable for the PI controllers, which represents one of the fundamental limitations of linear controllers.

Direct torque control

Direct torque control (DTC) is based on the estimation of torque and flux directly from the state variables of the AC machine. The torque and flux can be controlled by applying the suitable voltage vector, synthesized by the available switching states of the converter.

The required voltage vector is chosen via a switching table, as function of the actuation of the torque and flux loop hysteresis controllers (in terms of increasing or decreasing flux or torque for a certain operational point). Implementation of the DTC scheme for a multiphase AC drive is shown in **Figure 17**.

The main characteristics of DTC are its simple implementation and a fast dynamic response achieved by using hysteresis controllers. Furthermore, the required switching states are directly assigned from the switching table algorithm, so no modulator is needed.

Conclusions

In this chapter, the main advantages of multiphase machine drives for its application in high-demanding industrial processes, traction and renewable energy grid interfacing were presented, with their main focus on multiphase power converter topologies. Various technical issues related to classical multiphase converters, based on multicell arrangements, were discussed, as well as some new converter topologies, such as the nine-switch converter (NSC) and 11-switch converter (ESC), were introduced.

The main advantages of classical multiphase converter topologies, based on voltage source converters (VSC), are mostly referred to their topological simplicity and their capability to implement conventional sinusoidal-PWM-based techniques. On the other hand, the increasing number of semi-conductors and dc-link capacitors (in the case of multichannel arrangements) and the need for common mode current compensation are their major drawbacks. In this field of application, matrix converters arise as suitable alternative, which enables the possibility to handle multiple output phases. However, maximum voltage gain limitations and the complexity of the modulation and commutation strategies are their main disadvantages, when compared to classical topologies.

Nine-switch and eleven-switch converters appear as a middle-point alternative between multicell and matrix converter topologies. These topologies allow the use of sinusoidal-PWM-based modulation techniques, without the need of complex modulation and commutation strategies (as in the case of the matrix converter), using a single dc-link stage.

Implementation of control strategies for multiphase drive, such as field-oriented control (FOC) and direct torque control (DTC), has been easily achieved by using multi-space decomposition of the each state space vectors (representing a state space variable).

Author details

Carlos A. Reusser

Department of Electronics, Universidad Tecnica Federico Santa Maria, Valparaiso, Chile

*Address all correspondence to: carlos.reusser@usm.cl

References

[1] Levi E. Multiphase electric machines for variable-speed applications. IEEE Transactions on Industrial Electronics. 2008;55(5):1893-1909

[2] Levi E. Advances in converter control and innovative exploitation of additional degrees of freedom for multiphase machines. IEEE Transactions on Industrial Electronics. 2016;63(1):433-448

[3] Zhang Z, Matveev A, Ovrebo S, Nilssen R, Nysveen A. State of the art in generator technology for offshore wind energy conversion systems. In: Electric Machines Drives Conference (IEMDC), 2011 IEEE International; May 2011. pp. 1131-1136

[4] Li H, Chen Z. Design optimization and evaluation of different wind generator systems. In: International Conference on Electrical Machines and Systems, 2008. ICEMS 2008; Oct 2008. pp. 2396-2401

[5] Levi E, Bojoi R, Profumo F, Toliyat HA, Williamson S. Multiphase induction motor drives—A technology status review. IET Electric Power Applications. 2007;1(4):489-516

[6] Banerjee A, Tomovich MS, Leeb SB, Kirtley JL. Control architecture for a doubly-fed induction machine propulsion drive. In: 2013 Twenty- Eighth Annual IEEE Applied Power Electronics Conference and Exposition (APEC); March 2013. pp. 1522-1529

[7] Leon JI, Kouro S, Franquelo LG, Rodriguez J, Wu B. The essential role and the continuous evolution of modulation techniques for voltage- source inverters in the past, present, and future power electronics. IEEE Transactions on Industrial Electronics. 2016;63(5):2688-2701

[8] Venturini M, Alesina A. The generalised transformer: A new bidirectional, sinusoidal wave- form frequency converter with continuously adjustable input power factor. In: 1980 IEEE Power Electronics Specialists Conference; June 1980. pp. 242-252

[9] Lee MY, Wheeler P, Klumpner C. Space-vector modulated multilevel matrix converter. IEEE Trans-actions on Industrial Electronics. 2010;57(10): 3385-3394

[10] Wheeler PW, Clare JC, Apap M, Lampard D, Pickering SJ, Bradley KJ, et al. An integrated 30kw matrix converter based induction motor drive. In: 2005 IEEE 36th Power Electronics Specialists Conference; June 2005. pp. 2390-2395

[11] Abebe R, Vakil G, Calzo GL, Cox T, Lambert S, Johnson M, et al. Integrated motor drives: State of the art and future trends. IET Electric Power Applications. 2016;10(8):757-771

[12] Pena R, Cardenas R, Reyes E, Clare J, Wheeler P. A topology for multiple generation system with doubly fed induction machines and indirect matrix converter. IEEE Transactions on Industrial Electronics. 2009;56(10): 4181-4193

[13] Liu C, Wu B, Zargari N, Xu D, Wang J. A novel three-phase three-leg ac/ac converter using nine igbts. IEEE Transactions on Power Electronics. 2009;24(5):1151-1160

[14] Hava A, Kerkman R, Lipo T. Simple analytical and graphical tools for carrier based pwm methods. In: Power Electronics Specialists Conference, 1997. PESC '97 Record, 28th Annual IEEE, Vol. 2; Jun 1997. pp. 1462-1471

[15] Reusser CA, Kouro S, Cardenas R. Dual three-phase pmsg based wind energy conversion system using 9- switch dual converter. In: 2015 IEEE Energy Conversion Congress and Exposition (ECCE); Sep. 2015. pp. 1021-1022

[16] Reusser CA. Full-electric ship propulsion, based on a dual nine-switch inverter topology for dual three-phase induction motor drive. In: 2016 IEEE Transportation Electrification Conference and Expo (ITEC); June 2016. pp. 1-6

[17] Kumar EA, Satyanarayanat G. An eleven-switch inverter topology supplying two loads for common-mode voltage mitigation. In: 2017 IEEE PES Asia-Pacific Power and Energy Engineering Conference (APPEEC); Nov 2017. pp. 1-6

5

PV Plant Influence on Distribution Grid in Terms of Power Quality Considering Hosting Capacity of the Grid

Ivan Ramljak and Drago Bago

Abstract

Photovoltaic plants penetrate rapidly in distribution grid. Problems with their integration in distribution grid can exist in terms of load flow, protection settings, power quality, etc. This chapter analyzes influence of photovoltaic plants connection in distribution grid (0.4 and 10 kV voltage level) on power quality. The main focus will be on influence of photovoltaic plant connection point on distribution grid (hosting capacity—strength of the grid) in terms of power quality. Norms and regulations about influence of photovoltaic plants on distribution grid in terms of power quality will be analyzed. Influence of photovoltaic plants on distribution grid in theoretical aspects will be presented. Several case studies then will be described. Those case studies present different connection points of photovoltaic plants on distribution grid. Comparison of theoretical assumptions and real case studies will be compared. Some observations of real case studies and their impact on theoretical aspects, norms, and regulations about photovoltaic plant influence on distribution grid will be introduced.

Keywords: photovoltaic plant, distribution grid, power quality, hosting capacity, connection point, legal regulations

Introduction

Renewable sources (RES) (or distributed generation (DG) or embedded generation (EG)) are today one of the most actual fields in power engineering area. RES are very important factor in areas of environment protection, development of new technologies and smart grids. In the last several years, there is high penetration of RES on power grid regardless of voltage level. Those RES are with different technologies, but with same assignment: generation of renewable electrical energy. Considering source for electrical energy generation, RES can be based on sun radiation, wind power, waste utilization, exploitation of hydropower, etc. Rated power span of power plants based on RES can vary from 5 kW (microgeneration) to 150 MW (large generation) [1]. Nonrenewable sources (thermal and gas power plants and nuclear power plants) generally are connected on high voltage (HV) grid. It is due to their great rated power. In terms of that, middle- and low-voltage (MV and LV) distribution grid is generally without nonrenewable sources presence.

Without RES, distribution grid is a passive grid which transfers electrical energy from HV grid to consumers. But, penetration of RES makes distribution grid active. It is due to RES presence on all voltage levels (from HV to LV). That is due to great span of RES rated power. It means that RES can have connection in LV grid (e.g., PV plant with 10 kW) and in HV grid too (e.g., windfarm with more than 100 MW of rated power). Simply, conclusion can be derived. It is that conception of power system with RES penetration is changed. In area of electrical energy generation, RES today have great share in total electrical energy generation. But, electrical energy from, for example, PV plants and wind farms is uncontrollable and still difficult to foresee. That fact increased effort of researchers in area of storage systems development. RES created the most changes in distribution grid. Distribution grid, as already stated, became active grid. That fact made changes in distribution grid in terms of load flow conception, system protection, power quality, etc.

Those changes are challenge for distribution system operators (DSOs). Distribution system operation and management becomes very difficult for those DSO. A lot of research about RES penetration in distribution grid is carried out. The results are presented in various publications. Some observations can be found in [2–7]. References [3, 4] give fundamental knowledge about RES penetration in distribution grid. Basics, in terms of grid connection, power quality, system protection, economy, and reliability concepts, are described. In [4], influence of RES on power quality and system stability was analyzed. The conditions for RES connecting were described. The authors showed that carefully selected and placed DG can have positive impact on distribution grid in terms of improving voltage profile and system losses. In [5], technical requirements for the connection of the new DG resources on the distribution grid have been analyzed. Some experiences from Greek grid were presented. In [6], technical requirements and assessment criteria for RES connection on distribution grid for power quality and system protection- related issues are presented. Reference [7] introduces statistics modeling for planning purposes of RES integration in distribution grid. One of the greatest challenges in RES connection on distribution grid is in terms of power quality. Simply, before RES penetration in distribution grid, voltage drop was the highlighted problem in distribution grids. With RES, opposite, voltage increasing is the most prominent problem in distribution grids than voltage drop.

This is only one of the evidences of conception change in today's distribution grid. RES impact on power quality in distribution grid is analyzed in [2, 8–12]. Reference [8] analyzes short circuit power level with respect to power quality. Paper proposes a methodology to correlate rapid voltage changes to the load variations and to the short circuit power in point of common coupling (PCC). Presented analysis is valid for MV and high voltage (HV) grid. In [9], network impedance influence on power quality results in PCC is analyzed. Assessment procedures for connection of RES on distribution grid are presented. Registration and processing of power quality data in power grid with integrated RES is presented in [10]. In [11, 12], power quality monitoring and classification methods in distribution grids with penetrated RES are presented.

There are various power quality parameters. RES penetration in distribution grid influences the most on voltage quality parameters. The impact of RES on distribution grid among voltage quality parameters such as harmonics, flicker, and voltage magnitude is presented in [4–6]. In [9], analyzed voltage parameters are voltage changes, flickers, harmonics, and commutation notches. Reference [11] analyzes voltage fluctuations and harmonics. In [12], frequency, magnitude of voltage, flicker, unbalance, and harmonics are analyzed. In [13], harmonic, unbalance, and flicker are presented. The most analyzed voltage quality parameters, among literature, on which RES can have influence on distribution grid are voltage variations, flicker, and harmonics. As the share of RES increases, the development and use of legal regulations becomes very important factor for proper function of distribution grid.

Those legal regulations depend usually on country, due to each country specifics. But, there are some common international regulations. The best known is EN 50160, made by CENELEC in the middle 1990s. The other one is IEC 61000 series regulation on power quality. The IEEE regulations are present too in engineering practice, but in Europe, IEEE regulations usually cannot have legal status. In [3], EN 50160, IEC 61000 series, and some IEEE Standards are mentioned. PV plants are power sources that rapidly penetrate in power system. Generally, they use energy of the Sun and convert it to electrical energy [2]. PV plants are present on all voltage levels, from LV (10 kW rated power) to HV (more than 100 MW rated power).

This book chapter deals with influence of PV plant connection on distribution grid in terms of power quality. This chapter puts focus on dependence between strength of the grid (connection point of PV plant in the distribution grid) and PV plant rated power. In this chapter, theoretical assumptions will be analyzed. Further on, in order to reconsider theoretical assumptions, case studies will be presented too.

The chapter structure is as follows:

a. Influence of PV plant on power quality in connection point (PCC). Literature review (theoretical assumptions) and validation of it on case study.

b. Legal regulations of power quality considering RES influence on distribution grid. Literature review and current state.

c. Hosting capacity of the grid (strength of the grid) considering rated power of PV plant. Literature review and multiple case studies.

d. Comparison of theoretical and practical results of PV plant influence on distribution grid. Proposal of theoretical calculation on PV plant influence on distribution grid considering practical (measured) results. Verification on case study.

First, review of already known facts in this area will be given. Further, new conclusions will be introduced. The idea of further work will be presented based on new findings. At the end of the chapter, summary conclusions will be given.

In this chapter, power quality presents voltage quality parameters.

Influence of photovoltaic plant on power quality

PV plants can have connection in distribution grid in various PCC. That PCC potentially can be every bus in distribution grid. Generally, influence of connected PV plant in PCC (in terms of power quality) will be presented. This influence is analyzed among various literature. In [14, 15], influence of PV plants on waveform distortion in PCC was analyzed. It is concluded that if size of PV system is relatively low with respect to the short circuit power of the grid, there is no significant influence on the grid voltage quality. But, if short circuit power is relatively low, regarding PV system rated power, some problems considering power quality can be observed. According to [16], PV plants should not seriously degrade quality of supply with regard to harmonics. In [17], measurements in LV grid in PCC of PV plant connection are performed. Conclusion is that influence of PV plant in PCC is marginal in case of flicker and harmonics. This statement is valid for each measured point in grid (not only PCC). Same results for harmonics are obtained in [18].

Further on, in [19], PV plant connection on MV grid is analyzed. The result is that connected PV plant has positive contribution to the reduction of harmonic voltages distortion (with respect to the harmonic distortion occurring without connected PV plant). Power quality problems with connected PV plants are more often in rural grids (then in urban grid) due to usually longer feeders and small cross section of conductors [20]. In [21], this is presented as a potential problem too. Limitations of PV plants penetration in distribution grids apply only in weak grids [22]. But, PV plant penetration in weak grids can cause even better performance of those grids (improving voltage profile). Each case should consider individually, as stated in [22]. Simulations in [23] with PV plant connection in distribution grid showed no violation in the harmonic values. According to [24], in period of low irradiance, harmonic distortion is greater than in period with high irradiance.

Case study

Considering literature review, case study results will be presented. The goal is to compare theoretical assumptions and real case study results. A measurement of power quality for one PV plant was performed. Measurement lasted a week without PV plant and a week with connected PV plant, at PCC. Measuring (periods) was performed according to EN 50160:2011 issued by CENELEC (later analyzed). The most interesting voltage quality parameters will be presented:

- voltage unbalance—u (%);

- total harmonic distortion of voltage—THDU (%);

- slow voltage variations—U (V); and

- long-term flicker—P_{lt}.

The question about influence of PV plant on voltage quality in PCC is imposed. Rated power of PV plant is 23 kVA. Short circuit power of grid at PCC is 0.95 MVA. The PV plant is located at rural area connected to LV grid, on feeder with insulated overhead lines. The grid is in good condition (nearly reconstructed). There are two three-phase inverters in this PV plant. The goal is to find influence of PV plant connection on voltage quality at PCC. Analyzed results were in scope of 100% of measuring time periods. It is mandatory condition, so that total influence of PV plant on PCC in distribution grid can be seen.

Voltage unbalance

Figure 1 presents voltage unbalance (%) before and after PV plant connection. Due to clearness, 1-day results (24 hours, 10 min periods) of voltage unbalance before/after PV plant connection are presented in **Figure 1**. Similar is for all week too.

It is quite clear that any correlation of unbalance before and after PV plant connection cannot be found. Maximum (max.) and minimum (min.) value of unbalance is quite similar before (b.c) and after PV plant connection (a.c). It is shown in **Table 1** (for all week period).

From **Table 1**, it is visible that unbalance has lower value after PV plant connection!

Total harmonic distortion of voltage (THDU, %)

Figure 2 presents THDU values for L1 phase regarding P/P_n (%) values of PV plant. P/P_n presents ratio of generating power (P) and rated power (P_n) of PV plant. Due to clearness, 1-day results (24 hours, 10 min periods) of THDU (%) before/after PV plant connection are presented in **Figure 1**. Generally, results are similar for each day. Results for other two phases are similar.

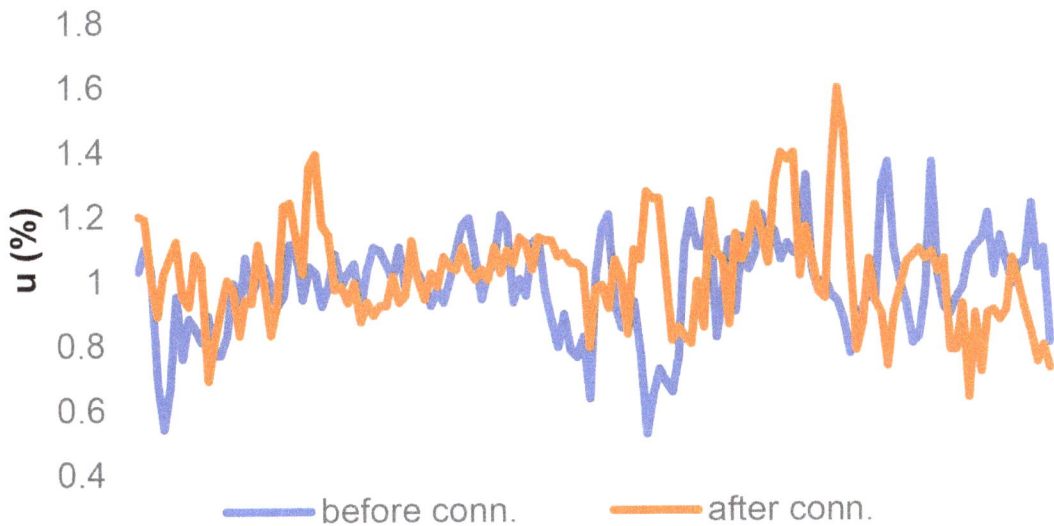

Figure 1. *Unbalance (%) before and after PV plant connection.*

Table 1. *Unbalance (%) values before and after PV plant connection.*

		b.c	a.c
u (%)	Maximum	1.98	1.60
	Minimum	0.52	0.50

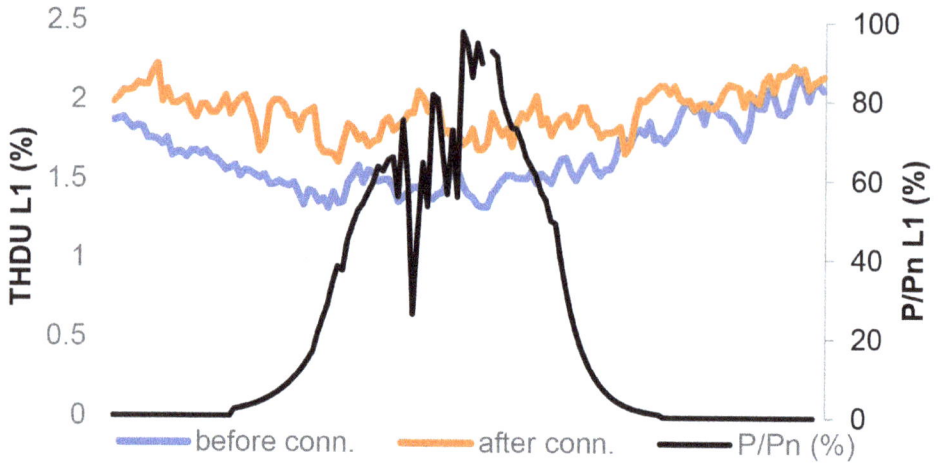

Figure 2. *THDU (%) for L1 phase before and after PV plant connection.*

It is quite obvious that after PV plant connection, THDU values are greater. Those increasing in THDU are the most presented mainly while PV plant produces electrical energy. But, analyzing 10 min values for all week period (all three phases), it is visible that max. and min. values of THDU are quite similar before (b.c) and after PV plant connection (a.c). It is shown in Table 2.

Table 2.*THDU (%) values before and after PV plant connection.*

		b.c	a.c
THDU (%)	Maximum	2.67	2.70
	Minimum	1.15	1.25

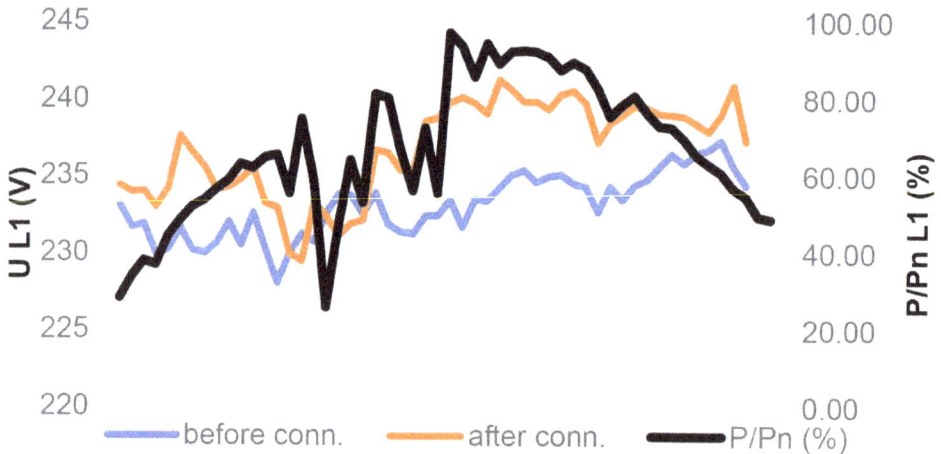

Figure 3. *Voltage values (V) for L1 phase before and after PV plant connection.*

Both (min. and max.) values are slightly increased after PV plant connection in this case. It is mainly in period when PV plant produces electrical energy. In percent value, for a week period, it is negligible. Generally, THDU values before and after PV plant connection are in scope of permitted values regarding any regulation.

Slow voltage variations

Voltage values in V for L1 phase, compared to P/P_n value, are presented in **Figure 3**. Those voltage values are 10 min interval for one PV plant generation day, between 8.00 and 17.00 hours. It is clear that PV plant generation increases voltage. As P/P_n is closer to 100%, voltage difference in V increases considering period before and after PV plant connection.

But, analyzing 10 min values for all week period (in all three phases), it is visible that max. and min. values of voltage values in V are quite similar before (b.c) and after PV plant connection (a.c). It is shown in Table 3. In this case, both (min. and max.) values are slightly increased after PV plant connection. In percent value, for a week period, that increase looks negligible. But, 246.36 V can be potentially high voltage, depending on regulations!

Figure 4 presents voltage values of phase L2 in V after PV plant connection, considering P/P_n (%) values. Those values present 10 min interval for 1 day of PV plant generation, between 8.00 and 17.00 hours. It is obvious that voltage value in V is related to P/P_n (%) ratio. Generally, greater P/P_n (%) ratio means greater voltage in V (voltage increase).

Long-term flicker

Figure 5 presents P_{lt} values before and after PV plant connection. On **Figure 5** is also represented current I (A) of PV plant phase L3 (I_{L3}). Measurement period is 10 min, from 20:00 to 20:00 (24 hours). In night period, flicker values are practically the same with/without PV plant. At day period, those values are greater with/without PV plant. There was not found any correlation between PV plant energy generation and flicker values. Table 4 presents P_{lt} values in PCC before/after PV plant connection for period of 2 weeks.

Table 3. *Voltage values in V before and after PV plant connection.*

		b.c	a.c
U (V)	Maximum	243.17	246.36
	Minimum	225.56	227.19

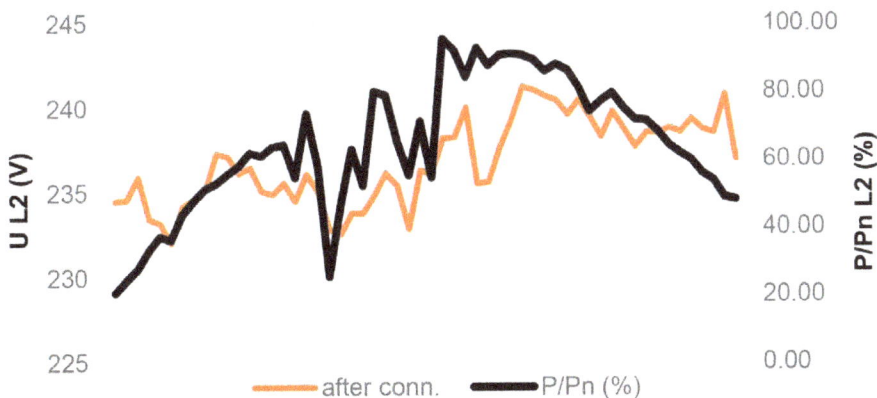

Figure 4. *Voltage values (V) for L2 phase considering P/P_n (%) values.*

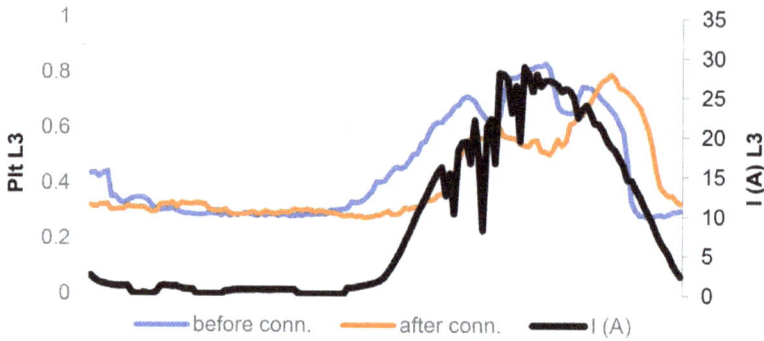

Figure 5. *P_{lt} L3 values before and after PV plant period regarding I_{L3} in A.*

Table 4. *P_{lt} values before and after PV plant connection.*

		b.c	a.c
P_{lt}	Maximum	1.391	1.244
	Minimum	/	/

It is obvious that PV plant does not contribute to increasing of P_{lt} at PCC. P_{lt} values have max. value before PV plant connection (not after connection) for this case study.

General conclusions of PV plant connection for case study

Some general conclusions for PV plant influence on voltage quality at PCC can be stated:

- Unbalance is quite similar before and after PV plant connection. From **Table 1** is visible that max. 10 min value of unbalance has lower value after PV plant connection. There is no correlation between unbalance values and PV plant generation.

- THD values are slightly increased after PV plant connection in this case. It is mainly in period when PV plant produces electrical energy. In percent value, for a week period, it is negligibly increasing. But, PV plant did have influence on THDU increasing (even slightly).

- It is obvious that voltage value in V is related to P/P_n (%) ratio. Generally, greater P/P_n (%) ratio means greater voltage in V (voltage increasing). PV plant has influence on voltage increasing due to its electrical energy generation.

- It is obvious that PV plant does not contribute to increasing of P_{lt} at PCC. P_{lt} values in this case have their max. before PV plant connection.

Legal regulations of power quality

Power quality is regulated with norms/standards. But, this area is not unique considering regulations. For this chapter, some of the most exploited norms/standards will be analyzed. Most used international norms/standards are as follows:

- EN 50160 by CENELEC,

- IEC standards, and

- IEEE standards.

These standards usually apply to normal weather conditions and exclude situations like storms, etc. Some of the most used norms/standards in terms of power quality in distribution grids will be shortly presented below. But, most of the countries exist with some legal regulations, like grid codes, in which voltage quality is exploited too.

EN 50160

In Europe, the most mentioned standard is EN 50160, published by CENELEC (European Committee for Electrotechnical Standardization). First published version of this norm is from the middle 1990s. It is probably the first norm dealing with voltage quality problems. This norm gives quantitative characteristics of voltage.

The norm describes characteristics of voltage parameters regarding permissible deviations. EN 50160 analyzes voltage characteristics considering frequency, wave- form, symmetry, and magnitude. Measurement of voltage must be performed for 7 days. Instruments for power quality measurement are described by IEC 61000-4- 30 standard. Procedure for power quality parameters measuring, according to EN 50160, is defined by IEC 61000-4-x standards. This norm gets a new issue every few years. Voltage quality limits, interesting for this analysis, are presented in Table 5. In Table 5, compliance limits for statistical evaluation of 95% of time are shown.

Table 5. *EN 50160—voltage quality parameters.*

Supply voltage parameter	Compliance limit
Voltage variations	±10%
Voltage unbalance	<2%
Long-term flicker	≤1
Harmonic voltage (THDU)	<8%

Table 6. *IEEE—voltage quality parameters.*

Supply voltage parameter	Compliance limit
Long-term flicker	≤0.8
Harmonic voltage (THDU)	<5%

IEEE standards

There is set of standards that consider voltage quality area, issued by IEEE (Institute of Electrical and Electronics Engineers). List of those standards can partly be found in [25]. Some voltage quality limits (available to authors) in scope for this analysis are presented in Table 6. In Table 6, compliance limits for statistical evaluation of 95% of time are shown. Standard that analyzes flicker is IEEE Std. 1453 and standard that analyzes harmonics is IEEE Std. 519.

IEC standards

IEC standards are issued by International Electrotechnical Commission. In scope of voltage quality are IEC 61000 Series Standards. Some voltage quality limits prescribed by IEC (available to authors) are shown in Table 7.

Now, short overview considering Tables 5–7 can be done. It is interesting that all three published standards have some unity elements. Mainly, 95% of time, voltage parameters must be in limited area. Slow voltage variations are usually limited 10% above/below nominal value. Unbalance is limited to 2%. P_{lt} is limited to 0.8 or 1. The greatest difference is, however, in THDU limits. The limited value varies from 5 to 8%.

Regarding the above-analyzed voltage quality parameters, voltage variations are the most exploited parameter. This is due to increased sensitivity of consumers on voltage values. It is a parameter of voltage quality on which consumers usually have the most complaints. Unfortunately, RES can potentially create even greater problems. Beside norms, in most of the countries exist local regulations (grid codes or similar) about permissible voltage values considering voltage level. Some of them are similar to the above-described norms, but some not.

Those regulations, based on internal country documents, can be divided into two classes. First regulation class presents permissible voltage values, which should be considered during distribution grid planning. Second regulation class considers limited voltage values during distribution grid exploitation. According to [3], per- missible voltage values in LV grid are -10 to +6% of nominal value in France. In MV grid, this value is -5 to +5%. In [3] is stated that in the United States, for MV grid, permissible voltage values are from -8 to +8% of nominal value. For Canada, those values are from -6 to +6%. It is interesting according to [3], that in Brazil, in LV grid, permissible voltage variations are from -14 to +6%. The examples below are for the case of grid planning. Below, some regulations for distribution grid in exploitation will be presented. For LV customers in Hungary, 95% of the 10 min voltages should be within 7.5% of the nominal voltage value, 100% of the 10 min voltages should be within 10% nominal voltage value. Moreover, 100% of the 1 min voltages should be within 15% nominal voltage value [3]. In Norway, 100% of the 1 min voltage values should be within 10% of the nominal voltage values [3]. For Spain, 95% of the 10 min voltage values should be within 7% of the nominal voltage value [3]. Medium voltage customers in France can sign a contract that gives them a voltage within 5% of a nominal voltage [3].

Table 7. *IEC 61000 series—voltage quality parameters.*

Supply voltage parameter	Compliance limit
Voltage variations	±10%
Long-term flicker	≤0.8
Voltage unbalance	<2%
Harmonic voltage (THDU)	THD < 8%

It is visible great difference between individual country regulations, regarding existence of norms issued by CENELEC, IEEE, and IEC.

Legal regulations of power quality and RES

If one closely looks norms/standards mentioned above, the logical question can be asked. Where are RES in those regulations? Indeed, slow voltage variations are usually limited 10% above/below nominal value, no matter is or is not RES presented in PCC. Considering Section 2.1, this can be a potential problem. If voltage value is, for example, +9% above nominal value, PV plant connection can potentially contribute to voltage increasing of, for example, +2%. It means that now voltage increasing is 11% in that PCC—out of limit. On the other side, if voltage value is, for example, +5% above nominal value, PV plant connection can potentially contribute to voltage increasing of, for example, +2%. It is now voltage increasing of 7%—permitted value. But, what if on the same feeder two more PV plants should be connected. This can lead to unallowed voltage increasing. This is valid for other voltage quality parameters. Norms/standards that describe influence of RES connection in PCC, in terms of voltage quality, are not present as such.

This area is usually controlled with regulations (documents), like grid codes, where influence of RES on voltage quality in electrical grid is defined. Those regulations vary between countries, regarding RES type, grid voltage level, current state of grid, etc. Literature review of permissible voltage quality deviation in PCC by RES connection is relatively modest.

Voltage increase after RES connection in PCC can be up to max. 3% [1, 4, 5]. P_{lt} value can be up to 0.65 according to [1] and 0.46 according to [4]. According to [13], compatibility level of P_{lt} in LV grid must be up to 0.8, and planning level of max. 0.4. P_{lt} must be up to 0.7 for MV grid [11]. Installed rated power of RES in LV grid can be up to 10% of rated power of transformer MV/LV [4]. It is clear that there is a wide literature approach about RES influence on voltage quality in distribution grid.

Hosting capacity of distribution grid and PV plant connection

Hosting capacity as a term can have different definitions. Considering voltage increasing, hosting capacity is max. permissible voltage increasing with connected DG, which causes max. voltage value strictly at the overvoltage limit. This leads to a conclusion that, considering voltage increasing, the hosting capacity is the amount of generation that gives a voltage increasing equal to the overvoltage margin [3].

The hosting capacity is the max. amount of generation that can be connected in some PCC without resulting in an unacceptable voltage quality or reliability for other customers. Hosting capacity can be defined as capacity of grid resistance on voltage quality changes due to RES connection in PCC. Distance between the distributed generator and the transformer is inversely proportional to the hosting capacity. The hosting capacity is smaller as distributed generator is further away from the transformer. The feeder size and length determine strongly hosting capacity of the grid [3].

Simply, hosting capacity of distribution grid presents strength of the grid. In other words, it presents capacity potential for RES connection in one PCC. Practically, it is not the same to connect one amount of RES rated power in urban distribution grid and on the end of long rural feeder made by small cross section of conductors.

On the other side, to be able to estimate the hosting capacity, the max. permissible voltage in grid should be known. Max. permissible voltage values are analyzed in previous subtitle.

It is very probably that greater amount of RES can be connected in urban LV grid than in rural LV grid, in PCC quite distant from transformer MV/LV. But, those are assumptions that should be verified by live case studies.

PV plant connection and hosting capacity of distribution grid: case studies

This section presents summarized results of voltage quality measurements in LV PCC for four PV plants. Considering hosting capacity of LV grid, PV plants are located in different places (in LV grid). Those grids are in different areas, with overhead lines and underground cables. It is all presented in **Figure 6**. PV plants are named with letters: A, B, C, and D.

PV plants A and C are placed in rural areas. PV plant A is placed not far from transformer MV/LV. PV plant C is placed at the end of the LV feeder. PV plants B and D are placed in urban areas. PV plant B is connected at LV buses of transformer station. PV plant D is set in city center near to transformer MV/LV.

Hosting capacity of LV grid in this case will be presented by three-phase short circuit power (S_{sc}) and PV plant rated power (S_r) ratio. Greater S_{sc}/S_r ratio should mean stronger grid—greater hosting capacity. In other words, as PCC is closer to transformer, greater distributed generator can be connected. Table 8 presents main characteristics of PV plants and related LV grids in PCC.

Table 9 presents summary of voltage quality results for analyzed PV plants. Measurements were performed a week before and after PV plant connection. Measurements are in accordance with EN 50160:2011. Analyzed voltage quality parameters include slow voltage variations, voltage unbalance, long-term flicker, and total harmonic distortion of voltage. Symbol Δ presents difference of max. values before/ after PV plant connection.

Before and after PV plant connection, max. values of voltage quality parameters are given. It is due to examination of worst practical cases.

Figure 7 presents voltage quality parameter results for PV plants, sorted by PCC weakness (S_{sc}/S_r). Lower S_{sc}/S_r value in PCC means weaker grid (smaller hosting capacity).

Voltage increase is proportional to weakness of LV grid. As shown in **Figure 7a**, as grid becomes stronger, voltage increase is lower (there can be decreasing too).

For PV plants B and D, voltage decrease has occurred. It means that voltage has lower values after PV plant connection and PV plant does not have any influence on voltage variations (values). The restriction is measurement period of only 1 week before and after PV plant connection. But, the probability of crucial result changing with longer measurement period is minor. For all other parameters such as Δu (%), ΔP_{lt}, and $\Delta THDU$, there is no correlation between results with/without PV plant and hosting capacity of the grid.

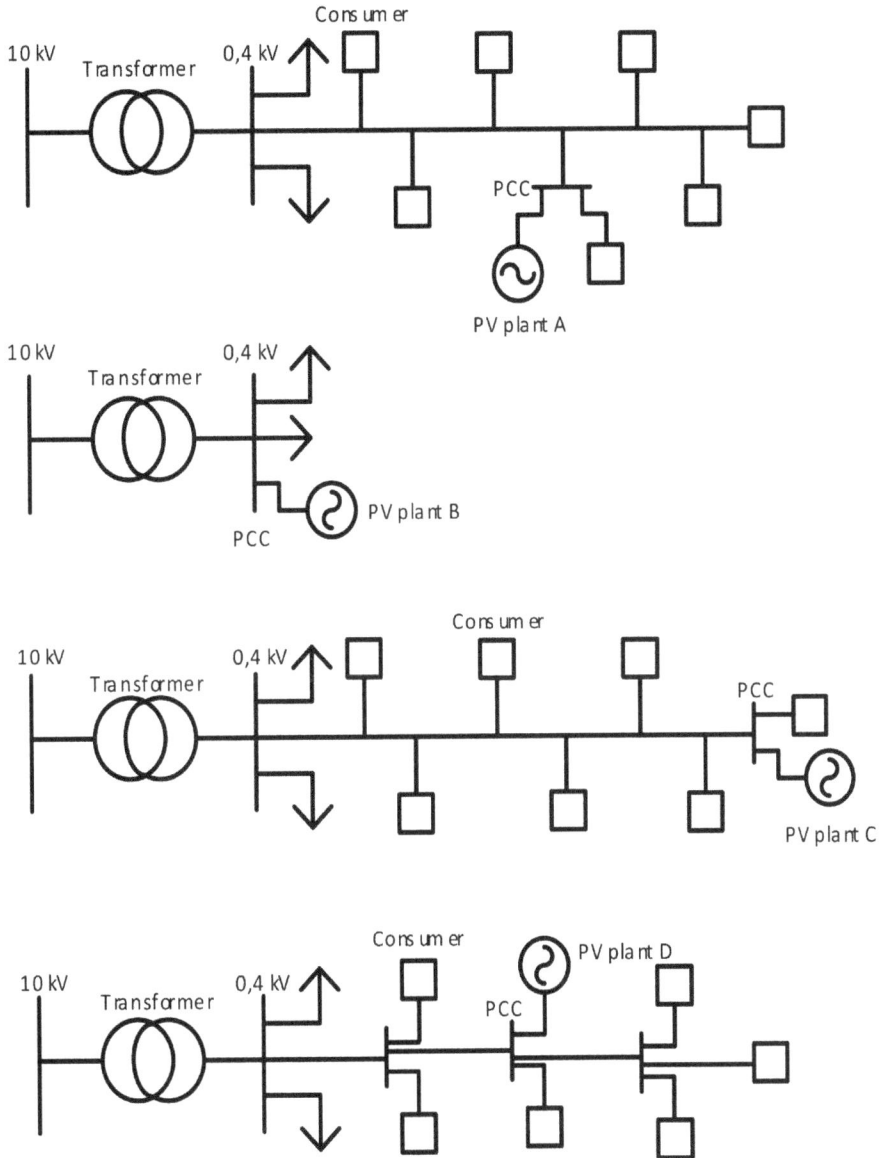

Figure 6. *Connection point of analyzed PV plants.*

Table 8. *Main characteristics of PV plants and related LV grids in PCC.*

	A	B	C	D
Sr (MVA)	0.023	0.085	0.023	0.023
Ssc (MVA)	0.95	8.85	0.21	10
Ssc/Sr	41.30	104.11	9.13	434.78

Conclusion is that strength of the grid has influence on voltage variations in distribution grid. It means that weak grid, regarding S_{sc}/S_r ratio, can lead to voltage quality problems in subjected distribution grid.

This cannot be concluded for other analyzed voltage quality parameters. **Figure 8** presents max. voltage values (voltage increase in % of rated voltage value) for PV plants before and after connection. Green and red lines present max. permissible voltage increase in PCC (in % of nominal value), assumed 5 and 7%. Those are realistic values in practice. If permitted max. voltage increase is 5%, there could be some technical problems. For PV plant B, even before PV plant connection, voltage values are increased. For PV plant A, voltage values are on margin. For PV plant C, after PV plant connection voltage values are increased. For permitted voltage increase of 7%, voltage values are in permitted area.

Table 9. *Summary of voltage quality results for analyzed PV plants.*

		A		**B**		**C**		**D**	
		b.c	a.c	b.c	a.c	b.c	a.c	b.c	a.c
U (%)	Maximum	4.99	5.11	5.52	5.32	4.03	6.45	1.4	1.04
	Δ	0.12		−0.2		2.42		−0.36	
u (%)	Maximum	1.33	1.34	0.45	0.48	1.23	1.26	0.82	0.87
	Δ	0.01		0.03		0.03		0.05	
P_{lt}	Maximum	0.94	0.98	0.92	0.61	3.45	4.15	1.16	1.17
	Δ	0.04		−0.31		0.7		0.01	
THDU (%)	Maximum	2.67	2.7	2.65	2.7	1.81	1.85	2.39	2.41
	Δ	0.03		0.05		0.04		0.02	

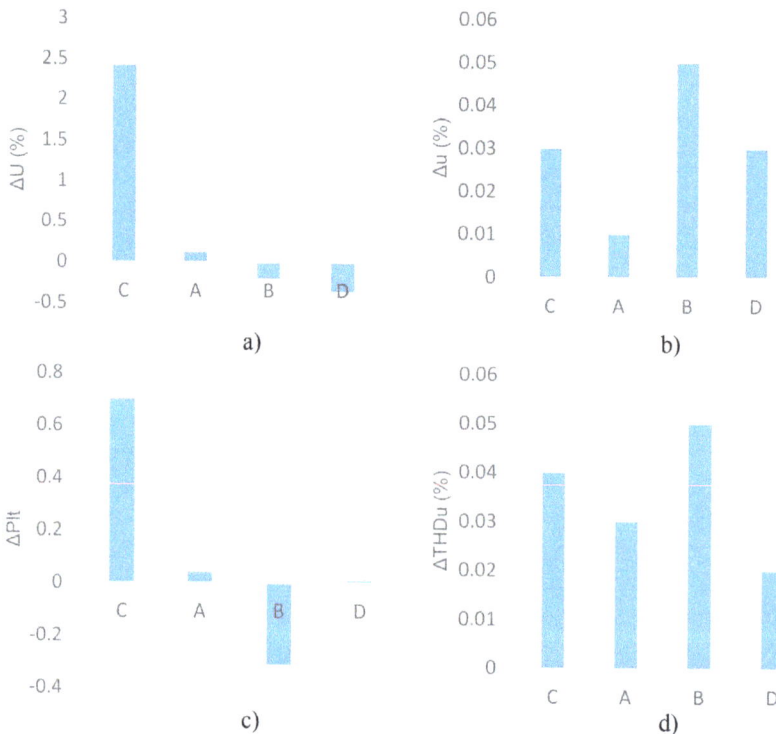

Figure 7. *Voltage quality parameter results for PV plants: (a) ΔU (%), (b) Δu (%), (c) ΔP$_{lt}$, and (d) ΔTHDU.*

Regarding voltage variations, **Figure 9** shows max. voltage value increase (ΔU (%)) in analyzed LV grids after PV plant connection. Permitted voltage value increase of 2 and 3% is assumed. It means that voltage increase after PV plant connection can be 2 or 3%. Those values are the most exploited in literature.

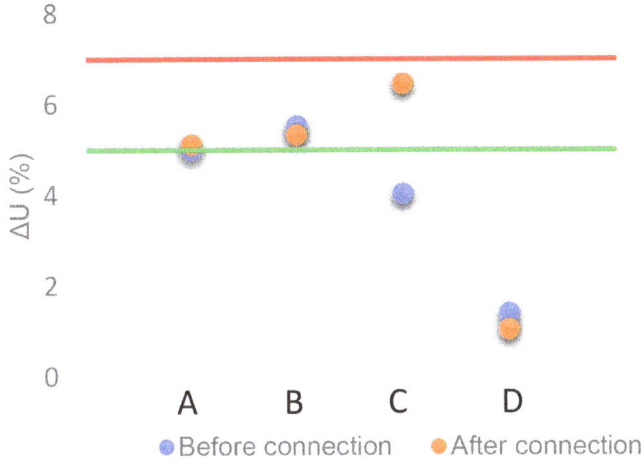

Figure 8. *Max. voltage values for PV plants before and after connection.*

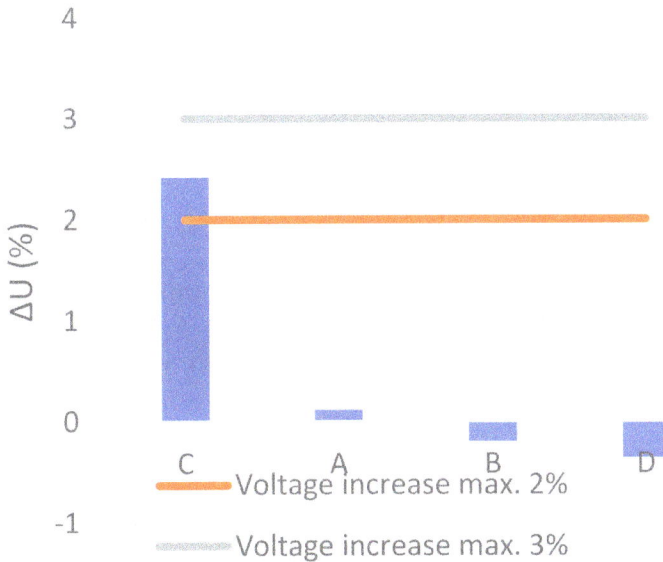

Figure 9. *Voltage increase in analyzed LV grids with connected PV plants.*

It means that if max. voltage increase in LV grid is 3%, PV plant C can operate. But, if max. voltage increase is 2%, PV plant cannot stay connected on the grid (some changes must be performed). Those changes should be in way of increasing short circuit current of the grid (greater hosting capacity of the grid) or in voltage regulations. Increasing of short circuit current in the grid means additional investment in grid (expending cross section of conductors or building new feeder for PV plant connection). Voltage regulation can be performed with inverters. This means regulation of power factor of PV plant (cos φ(P) or Q(U), for example). It is possible, but it usually means additional power losses in grid.

For example, let it be max. permitted voltage increase of 3% (considering nominal value) after PV plant connection. On the other side, let it be max. permitted voltage value increase of 5% (considering nominal value) in LV grid. It is obvious that PV plant B is out of permitted area for normal operation (**Figures 8 and 9**). But, even before PV plant connection, voltage values over permit value were increased. Moreover, after PV plant connection, voltage values were decreased. Another example is opposite. Let it be max. permitted voltage value increase of 2% for PV plant connection. On the other side, let it be max. permitted voltage value increase of 7% (considering nominal value) in LV grid. It is obvious that PV plant C is out of permitted area for normal operation (**Figures 8 and 9**). But, after PV plant connection, voltage values in PCC have permit value (regarding increase of voltage value more than 2% in PCC). This is area of sharing hosting capacity. It will be a challenge for further research.

General conclusions of PV plant connection regarding hosting capacity of LV distribution grid

Some conclusions for influence of PV plant on voltage quality at PCC, considering hosting capacity of that LV distribution grid, are as follows:

- PV plant has influence on voltage increase in PCC, considering hosting capacity of distribution grid. As the distribution grid is weaker, voltage increase is higher.

- THDU values, unbalance, and long-term flicker are apparently independent on hosting capacity of the grid.

- Contradiction is possible, in the way that voltage value increase after PV plant connection has permitted value, but in PCC, voltage value is unacceptable (increased); and in opposite, voltage value increase after PV plant connection has unacceptable value (increased voltage value), but voltage value in PCC is in limit values.

Comparison of theoretical and practical influence of PV plant connection in PCC considering slow voltage variations: case study

In this section, comparison of theoretical and practical results of PV plant influence in PCC considering slow voltage variations will be presented. This is important comparison, since in previous section, slow voltage variations (voltage increase) in PCC are turned out as potential problem for PV plant connection in distribution grid.

Practical results of measurements before and after PV plant connection are presented. A MV feeder (nominal voltage of 10 kV) with connected PV plant is observed. This PV plant will be named by letter E. Some basic parameters of PV plant and MV distribution grid are shown in **Table 10**.

Measurement site at 10 kV voltage level is shown in **Figure 10**. Theoretically, steady-state slow voltage variation (voltage increase) in PCC can be formulated as follows:

$$\Delta u(\%) = 100 \cdot \frac{S_r}{S_{SC}} \cdot \cos(\psi_k + \phi) \tag{1}$$

where ψ_k is phase angle of the network impedance and ϕ is the phase angle of RES output current. Eq. 1 is simplified, but accurate enough for practical purposes [5]. Worst case considering voltage increase in PCC is for cos ($\psi_k + \phi$) = 1. In this case study, result for voltage increase is **3.60%** (after PV plant connection) for cos ($\psi_k + \phi$) = 1.

Table 10. *Basic parameters of PV plant and MV distribution grid.*

	E
Sr (MVA)	0.920
Ssc (MVA)	25.46
Ssc/Sr	27.67

Figure 10. *Measurement site look for PV plant E.*

In **Figures 11** and **12**, practical measurement of line voltages (in kV) for a week before and after PV plant connection is shown.

Table 11 presents max. voltage values measured according to EN 50160:2011 (10 min value) before and after PV plant connection.

It is visible that voltage increase due to PV plant connection has max. value of 2.18%.

Max. voltage value increase after PV plant connection is close to 5%, that is, max. value for acceptable voltage value increase in MV grids according to [3].

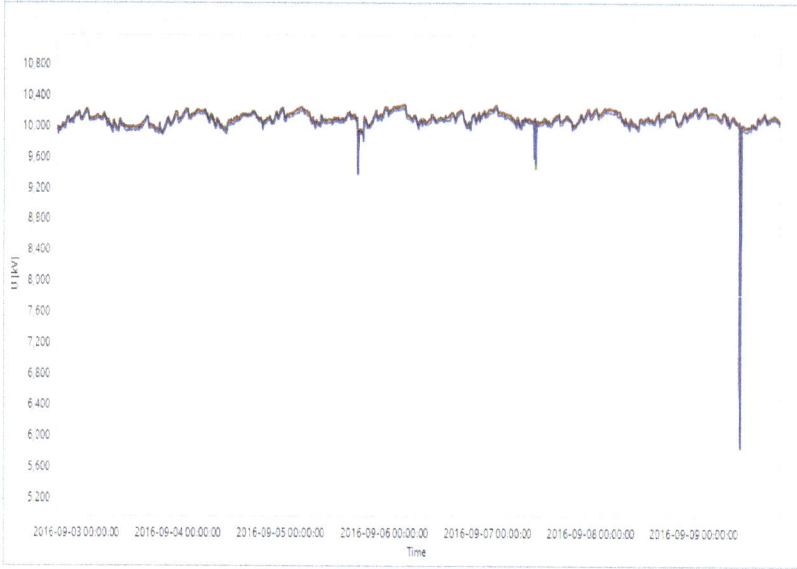

Figure 11. *Voltage measurement in PCC before PV plant connection.*

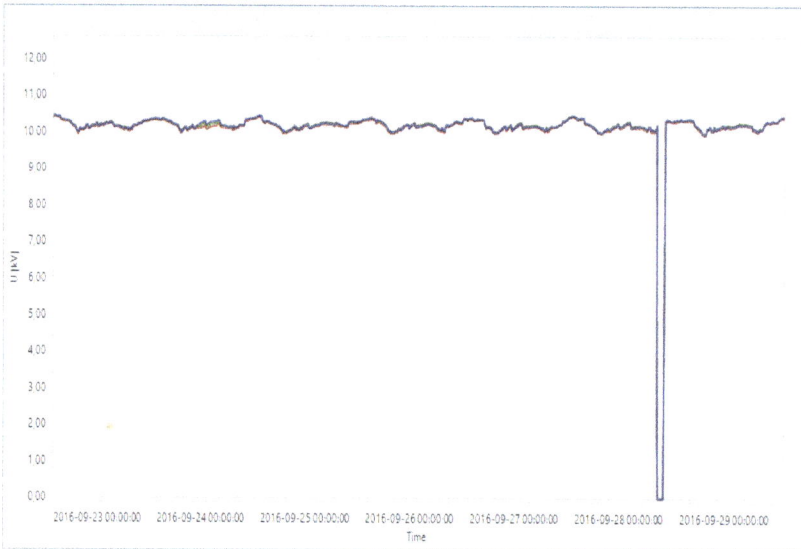

Figure 12. *Voltage measurement in PCC after PV plant connection.*

Table 11. *Slow voltage variation max. values before and after PV plant connection.*

		E	
		b.c	a.c
U (%)	Maximum	2.73	4.91
	Δ	2.18	

Comparing of theoretical and practical voltage value increase in PCC, after PV plant connection, can be performed. It is theoretical voltage increase of 3.60% comparing to 2.18% measured
in practice. It is obvious that theoretical voltage value increase is greater due to assumption cos

$(\psi_k + \phi) = 1$. This assumption obviously leads to max. voltage value increase, which is hard to achieve in practice.

So, if theoretical result is acceptable, there should not be any problem in practice! **Table 12** shows other voltage quality results (u, P_{lt}, THDU) before and after PV plant E connection.

Table 12. *Voltage quality parameters for PV plant E, before and after connection.*

		E	
		b.c	a.c
u (%)	Maximum	0.40	0.40
	Δ	0	
P_{lt}	Maximum	1.57	1.70
	Δ	0.13	
THDU (%)	Maximum	2.29	2.51
	Δ	0.22	

It is clear that for each showed parameter, there is no practically any difference in results before and after PV plant connection.

Conclusion

This chapter deals with PV plant influence on distribution grid. Literature review about PV plant influence on distribution grid is summarized. After that, case studies based on appropriate measurements are presented. The connection point of PV plant considering hosting capacity of distribution grid in terms of power quality was analyzed. Considering content of this chapter, following conclusions can be implemented:

- Power quality is regulated with some norms/standards. However, this area is not unique considering present regulations, which is presented in this chapter. There is lack of re-striction condition documents for RES connection on distribution grid. In most of the countries exist local legal regulations, like grid codes, in which voltage quality is exploit-ed in terms of RES influence on distribution grid.

- A case study of PV plant connection in LV grid is presented. The influence of PV plant on distribution grid in PCC, considering basic voltage quality parameters, was analyzed. Unbalance is quite similar before and after PV plant connection. THDU values are slightly increased after PV plant connection in this case. It is mainly in period when PV plant produces electrical energy. In percent value, for a week period of measurement, it is negligibly increasing.

- It is obvious that voltage value in V is related to P/P_n (%) ratio. Generally, greater P/P_n (%) ratio (greater generation of PV plant) means greater voltage value in volts (voltage increasing). PV plant has influence on voltage increasing due to its electrical energy generation. It was found that main potential problem for PV plant connection in distribution grid can be voltage increase.

- Finally, it is obvious that PV plant does not contribute to increasing of P_{lt} at PCC.

- A comparison of PV plants influence on distribution grid for different connection points of PV plants, considering strength of the grid, is presented. Strength of the grid is in fact hosting capacity of the grid. Four PV plants with different connection points (different S_{SC}/S_r ratio) were compared. PV plant connection point has influence on voltage increasing in PCC, considering hosting capacity of distribution grid. As the distribution grid is weaker, voltage increase is higher. THDU values, unbalance, and long-term flicker are apparently independent on hosting capacity of the grid. Contradiction is possible, in the way that voltage increase by PV plant connection has permitted value, but in PCC, voltage is increased. And in opposite, voltage increase by PV plant connection is increased, but voltage value in PCC is in limit values. This statement is valid for LV and MV grid. This area of contradiction falls in area of hosting capacity sharing and it is challenge for future work.

- Comparing of theoretical and practical voltage increase values in PCC was presented. Theoretical voltage value increase is greater due to the assumption $\cos(\psi_k + \phi) = 1$. This assumption obviously leads to max. voltage increase, which is hard to achieve in practice. So, if theoretical result is acceptable, there should not be any problem in practice. But, this analysis should not be only conducted analysis. It should be framework for more detailed analysis of PV plant influence on distribution grid. For each new PV plant connection, this analysis should be implemented.

Main contributions of this chapter can be summarized as follows:

- Proving of correlation of voltage value increase considering P/P_n (%) ratio increase. As the distribution grid is weaker (low S_{SC}/S_r ratio), voltage increase is higher.

- Voltage value increase in practice should not be greater than increase obtained by calculation considering assumption that $\cos(\psi_k + \phi) = 1$.

- PV plant does not contribute in increasing of P_{lt} and unbalance at PCC. Increasing of THDU in PCC is negligible after PV plant connection. These parameters are apparently independent of hosting capacity of the grid.

Acknowledgements

The authors would like to thank to JP Elektroprivreda HZ HB d.d Mostar for helping in issuing this chapter.

Author details

Ivan Ramljak* and Drago Bago

J. P. Elektroprivreda HZ HB d.d Mostar, Bosnia and Herzegovina

*Address all correspondence to: ivan.ramljak@ephzhb.ba

References

[1] Sikorski T, Rezmer J. Distributed generation and its impact on power quality in low-voltage distribution networks. In: Luszcz J, editor. Power Quality Issues in Distributed Generation. Rijeka, Croatia: IntechOpen; 2015. pp. 1-39. DOI: 10.5772/59895

[2] Jenkins N, Allan P, Crossley D, Kirschen D, Strbac G. Embedded Generation. 1st ed. London: The Institution of Engineering and Technology; 2000. 292 p. DOI: 10.1049/ PBPO031E

[3] Bollen M, Hassan F. Integration of Distributed Generation in the Power System. 1st ed. Chichester: Wiley; 2011. 528 p. ISBN: 978-1-118-02902-2

[4] Mgaya EV, Muller Z. The impact of connecting distributed generation to the distribution system. Acta Polytechnica. 2007;47:96-101. ISSN: 1805-2363

[5] Papathanassiou SA, Hatziargyriou ND. Technical requirements for the connection of dispersed generation to the grid. In: Proceedings of IEEE PES Summer Meeting; 15-19 July 2001; Canada. New York: IEEE; 2002. pp. 749-755

[6] Papathanassiou SA. A technical evaluation framework for the connection of DG to the distribution network. Electric Power Systems Research. 2007;77:24-34. DOI: 10.1016/j. epsr.2006.01.009

[7] Toubeau JF, Klonari V, Lobry J, De Greve Z, Vallee F. Planning tools for the integration of renewable energy sources. In: Cao W, Hu Y, editors. Renewable Energy. Rijeka, Croatia: IntechOpen; 2016. pp. 227-260. DOI: 10.5772/61758

[8] Allegranza V, Ardito A, De Berardinid E, Delfanti M, Lo Schiavo L. Assessment of Short Circuit Power Levels in HV and MV Networks with Respect to Power Quality [Internet]. 2007. Available from: http://www.cired. net/publications/cired2007/pdfs/ CIRED2007_0161_paper.pdf [Accessed: October 10, 2018]

[9] Meyer J, Ammeter U, Hanzlik J, Zierlinger J. Methods for the assessment of network disturbances in distribution networks. In: 3rd International Conference on Harmonics and Quality of Power; 28 September-1 October 2008; Australia. New York: IEEE; 2008. pp. 1-6

[10] Leonowicz Z, Rezmer J, Sikorski T, Szymanda J, Kotyla P. Wide-area system of registration and processing of power quality data in power grid with distributed generation: Part I. System description, functional tests and synchronous recordings. In: 14th International Conference on Environment and Electrical Engineering; 10-12 May 2014; Poland. New York: IEEE; 2005. pp. 175-181

[11] Golovanov N, Lazatoiu GC, Roscia M, Zaninelli D. Monitoring power quality in small scale renewable energy sources supplying distribution systems. In: Zobaa AF, editor. Power Quality Issues. Rijeka, Croatia: IntechOpen; 2013. pp. 227-260. DOI: 10.5772/53464

[12] Cobben S. Power quality monitoring and classification. In: Eberhard A, editor. Power Quality. Rijeka, Croatia: IntechOpen; 2011. pp. 103-128. DOI: 10.5772/595

[13] Hernandez JC, Ortega MJ, De la Cruz J, Vera D. Guidelines for the technical assessment of harmonic, flicker and unbalance emission limits for PV-distributed generation. Electric Power Systems Research. 2011;81: 1247-1257. DOI: 10.1016/j.epsr.2011. 03.012

[14] Chicco G, Schlabbach J, Spertino F. Experimental assessment of the waveform distortion in grid-connected photovoltaic installations. Solar Energy. 2009;83:1026-1039. DOI: 10.1016/j. solener.2009.01.005

[15] Ramljak I, Bago D. Influence of PV plant connection on voltage quality parameters considering connection point in distribution grid. In: Proceedings of the First International Colloquium on Smart Grid Metrology (SmaGriMet); 24-27 July 2018; Croatia. New York: IEEE; 2018. pp. 1-5

[16] Eltawil MA, Zhao Z. Grid-connected photovoltaic power systems: Technical and potential problems—A review. Renewable and Sustainable Energy Reviews. 2009;14:112-129. DOI: 10.1016/j.rser.2009.07.015

[17] Kopicka M, Ptacek M, Toman P. Analysis of the power quality and the impact of photovoltaic power plant operation on low-voltage distribution network. In: Proceedings of Electric Power Quality and Supply Reliability Conference; 11-13 June 2014; Estonia. New York: IEEE; 2014. pp. 99-102

[18] Kontogiannis KP, Vokas GA, Nanou S, Papathanassiou S. Power quality field measurements on PV inverters. International Journal of Advanced Research in Electrical, Electronics and Instrumentation Engineering;2013(2): 5301-5314. ISSN: 2278-8875

[19] Barbu V, Chicco G, Corona F, Golovanov N, Spertino F. Impact of a photovoltaic plant connected to the MV network on harmonic distortion: An experimental assessment. Electrical Engineering. 2013;75:179-193. ISSN: 2286-3540

[20] Canova A, Giaccone L, Spertino F, Tartaglia M. Electrical impact of photovoltaic plant in distributed network. In: Proceedings of IEEE Industry Application Annual Meeting; 23-27 September 2007; USA. New York: IEEE; 2007. pp. 1450-1455

[21] Ramljak I, Ramljak I. PV plant connection in urban and rural LV grid: Comparison of voltage quality results. In: 2nd ISPQ Symposium; 21-24 June 2018; Bosnia and Herzegovina. 2018. pp. 1-8

[22] Ivanovici TD, Ionel M, Dogaru- Ulieru V, Mihaescu S. The influence of photovoltaic systems on low-voltage grids. WSEAS Transactions on Environment and Development. 2011;3: 65-74. ISSN: 1790-5079

[23] Papaioannou IT, Bouhouras AS, Marinopoulus AG, Alexiadis MC, Demoulias CS, Labridis DP. Harmonic impact of small photovoltaic systems connected to the LV distribution network. In: Proceedings of 5th International Conference on the European Electricity Market; 28-30 May 2009; Portugal. New York: IEEE; 2008. pp. 1-6

[24] Patsalides M, Evagorou D, Makrides G, Achillides Z, Georghiou GE, Stavrou A, et al. The Effect of Solar Irradiance on the Power Quality Behaviour of Grid Connected Photovoltaic System [Internet]. 2007. Available from: http:// icrepq.com/icrepq07/284-patsalides.pdf [Accessed: September 15, 2018]

[25] Tokić A, Milardić V. Power Quality. 1st ed. Tuzla: PrintCom; 2015. 319 p. (in Croatian)

Perspectives on Dual-Purpose Smart Water Power Infrastructures for Households in Arid Regions

Dana Alghool, Noora Al-Khalfan, Stabrag Attiya and Farayi Musharavati

Abstract

In hot arid climates, freshwater and power are produced simultaneously through seawater desalination since these regions receive little rainfall. This results in a unique urban water/power cycle that often faces sustainability and resilience challenges. Elsewhere, such challenges have been addressed through smart grid technologies. This chapter explores opportunities and initiatives for implementing smart grid technologies at household level for a case study in Qatar. A functional dual-purpose smart water/power nanogrid is developed. The nanogrid includes multiloop systems for on-site water recycling and on-site power generation based on sustainability concepts. A prototype dual-purpose GSM-based smart water/ power nanogrid is assembled and tested in a laboratory. Results of case study implementation show that the proposed nanogrid can reduce energy and water consumptions at household level by 25 and 20%, respectively. Economic analysis shows that implementing the nanogrid at household level has a payback period of 10 years. Hence, larger-scale projects may improve investment paybacks. Extension of the nanogrid into a resilient communal microgrid and/or mesogrid is discussed based on the concept of energy semantics. The modularity of the nanogrid allows the design to be adapted for different scale applications. Perspectives on how the nanogrid can be expanded for large scale applications are outlined.

Keywords: water conservation, energy efficiency, smart water, smart grids, renewable energy, nanogrid, energy semantics

Introduction

Water and energy are among the most important commodities in life. They support growth, development, and human survival on earth. Consequently, sustainable water and energy supply have become critical issues of consideration in most parts of the world [1, 2]. Moreover, the water and energy nexus has been a great subject of debate for decades [3, 4]. For example, the United Nations has predicted a 40% global shortfall of water availability by 2030 and a 50% global short fall on energy [5, 6].

In spite of these observations, the demand for energy has been on the rise as various national economies become more and more advanced. In addition, climate change studies have projected unique changes in urban water cycles, thus making it more difficult for national economies to balance water supply and distribution now and in future plans. These difficulties add more strain on energy supply and freshwater access [7]. Freshwater concerns are even more critical in hot arid regions characterized by low rainfall and harsh climate. This chapter discusses potential solutions to sustainability challenges with reference to water and energy conservation. The underlying theme lies in that implementation of smart water and energy technologies has a significant impact on water and energy conservation practices in arid regions.

In most arid regions, water and energy supply networks are implemented as separate single-loop systems. This means that the stages of the current water/power cycle do not intersect and yet these commodities are produced simultaneously in dual-purpose water/power production plants. In practice, the urban water/power cycles often face challenges that require further investment by local authorities in a bid to mitigate the effects of sustainability challenges. In addition, population growths and rapid economic developments often strain water/power supply and distribution networks, thus making the urban water/power cycle less sustainable. It is therefore important to rethink the urban water/power cycle in a bid to develop water/power infrastructures that can improve the water/power use efficiencies at the household level by incorporating smart technologies.

Smart technologies have been implemented with benefits that support sustainability goals [8–10]. In the public literature, smart energy grids have been discussed thoroughly by many authors [11–13]. Smart energy grids have also been successfully implemented in various parts of the world [14–16]. A number of benefits associated with smart energy grids have been identified including economic [17, 18], environmental [18, 19], reliability [20, 21], and customer choice [22, 23]. Such benefits significantly contribute to both resilience and sustainability. While the literature is overwhelmed with smart energy grids, relatively little is known about smart water grids [24–26]. Of late, smart water grids have been found to hold a lot of potential for unlocking the requirements for a sustainable, stable, reliable, high-quality, resilient, and secure water supply system. Another prominent gap in the public literature lies in that water and energy smart grids are usually discussed separately. While this may be appropriate for other regions of the world, the unique connection between water and energy in arid regions requires special considerations and technologies that are more appropriate. In the Gulf Region, for example, water and energy are produced simultaneously [27–29]. It has been shown that there is an inherent link between energy and water [30–32]. It is, therefore, necessary to investigate opportunities and initiatives for developing dual-purpose water/power smart grids. Perspectives on the design and operations of such a smart grid will be discussed with reference to a case study in Qatar.

The common practice in Qatar is that once water and power are produced, they are distributed separately to residential areas. At the household level, water is pumped into a tank positioned at the roof of villas. After use at different end points, this water is directed into a sewer line where it is mixed with black water and further directed to wastewater reservoirs. Separating and reusing this water at the source (household) may prove beneficial. On the other hand, solar energy in Qatar is currently found in isolated areas that are far from grid connections. Most of the energy generated from solar is used as supplements to the main grid power. There is a need to increase the fraction of renewable energy in Qatar since the insolation is relatively high. This can position Qatar toward achieving sustainability goals as stipulated in the Qatar National Vision 2030.

One way of addressing sustainability issues is to closely examine the 6Rs of sustainability, i.e., reduce, reuse, rethink, recycle, refuse, and repair [33]. The purpose in implementing the 6Rs is to obtain the most practical benefits from products, processes, and systems and to generate the least amount of wastes. This approach also activates other external positive issues such as pollution reduction, resource saving, and avoidance of greenhouse gas emissions. The discussions in this chapter derive inspiration from four of these 6Rs, i.e., reduce, reuse, recycle, and rethink. *Rethink* is about trying to think (in a different way) how to generate electricity and provide use- ful water in order to minimize the consumption of the main grid electricity and fresh- water. For example, generating electricity from the velocity of clean or wastewater in water pipes (in-pipe hydropow- er generation) and treating the wasted water instead of disposing it to the main sewage directly after use are noble initiatives that can help in conserving both energy and water. In addition, current system designs for water are single-loop system from utilities. The idea in this work is to investigate the usefulness of multi-loops of water (freshwater, gray water, and black water) and energy (main grid supply, renewable energy micro-generation, and in-pipe hydroelectricity) at the household level in a bid to rethink and reuse available resources to the maximum possible. Design, development, and implementations of such multiple loops of energy and water deviate from the common single-loop systems and thus constitute an initiative for rethinking the energy and water networks at end-use locations.

Reduce is about reducing and minimizing the wasted water "produced" in the household as well as reducing the consumption and electricity from the main grid supply by implementing re- newable energy and smart technologies in the existing infrastructure. It is also about behavioral changes due to the conscious realization, by residents, of "wasteful" consumption of water and energy. *Reuse* is about reusing gray water produced in the house after treating it. The treated water can be "reused" for watering the gardens, car washing, as well as toilet flushing instead of "throwing the water down the drain." *Recycle* is about collecting the gray water that is produced at different end points in the house, such as sinks, showers, and washing machines for the pur- pose of treating and reusing the gray water at the source instead of sending it to the main sewage line where it is further contaminated by black water.

Based on the concepts discussed in the previous paragraphs, this chapter discusses the devel- opment of a smart dual-purpose water/power nanogrid under the climatic conditions in Qatar.

According to the Qatar National Vision 2030, Qatar aspires to be an advanced society capa- ble of sustaining its development and providing a high standard of living for its residents [34]. However, with the current population explosion and numerous construction projects, the utility companies in Qatar may face a number of water and electricity consumption challenges. For example, the residents in Qatar consume nearly twice the average consumption of water and electricity in other parts of the world, the EU being a specific example [35]. Statistical projections show that these consumption rates are expected to double in the near future, thus further straining the balance between water/energy supply and demand. In a bid to provide solutions for these challenges, the effects of implementing smart technologies are discussed in this chapter. A combination of smart water and smart energy technologies are discussed, and perspectives on how to integrate them into a functional nanogrid for a single household are outlined. The motivation emanates from the observation that residential water and energy infrastructures often waste substantial quantities of freshwater and energy. Therefore, there is a need to reduce water and energy consumptions at the household level.

Background

Water and energy resources are communally and reciprocally linked since meeting energy needs requires water and vice versa [3, 4]. The consensus is that saving water saves energy and energy efficiency opportunities are often linked to water savings. Albeit, both initiatives result in less carbon emissions. In hot arid climates, this relationship is intertwined since water and energy are produced simultaneously in dual-purpose water/power production plants. Therefore, addressing water and energy issues in tandem can result in significant benefits for utility companies.

Improving efficiency of energy and water in the supply and demand sides can allow national economies to reduce resource consumptions as well as maximize benefits for utilities, consumers, businesses, and communities. National economies need to increase water and energy security while reducing the environmental impacts of water and energy use. This means that available water and energy must be used more efficiently. Energy consumption in water reticulation systems can be reduced by using energy recovered from household water systems and wastewater at nanoscale to produce power on-site. Power consumption can be reduced at the household level by, for example, giving residents detailed energy consumption information that can be used by residents to decide on how best to use energy in their homes.

Assessment of end-use energy and water efficiencies provides information that can be used to find ways of reducing the strain on the main power grid and water distribution network. However, a number of barriers and challenges may exist. In the Gulf countries, for example, there is currently an overall trend toward larger homes and a greater variety of appliances and electronics in each home. This trend further strains the water and energy resources at the national level and hence contributes to the imbalance on water and electricity supply and demand. Options for increasing end-use energy efficiencies include renewable on-site power generation, implementing well-designed energy codes and standards, improving end-use appliance energy efficiency, using efficient plumbing fixtures, and educating homeowners about behavioral changes that will result in significant reductions in energy consumptions. Since water and power are produced in expensive seawater desalination plants, water conservation and water recycling are important initiatives that can be used to leverage end-use efficiencies. Furthermore, such initiatives support sustainable developments.

Energy use in residential buildings account for about 17% of US greenhouse gas (GHG) emissions [36, 38]. Unlike the Gulf countries, the large share of residential building energy consumption is attributable to space heating and cooling, which varies with climate conditions. In the Gulf countries, cooling accounts for about 70% of energy used in residential buildings [37]. Other energy uses are related to providing power to various household appliances that are used randomly. Reducing energy consumptions of these end uses is difficult since it requires different technological improvements for each appliance as well as behavioral changes that aim at increasing energy efficiency and conservation . This represents significant challenges to sustainability goals.

While many options are available for providing clean water, seawater desalination has taken the center stage in the Gulf countries. Common technologies for sea- water desalination include multistage flash distillation, multi-effect distillation, and reverse osmosis. Since environmental concerns are on the rise, renewable energy technologies are becoming more important and attractive partners for powering water desalination projects in arid regions [39], while desirable, renewable energy cannot cope with the quick, discontinuous, and uncontrollable falls and peaks in electricity demand. Since renewable energy technologies depend on the season and the time of the day, their integration poses challenges to the traditional grid systems. Generating electricity from renewable energy, mainly photovoltaics (PV), wave, and wind power depend extremely on the unpredictable nature of weather conditions and status [40]. If new electric devices are employed in the renewable energy-based electricity systems, great achievements can be realized. Examples of electric devices and components that can support renewable energy electricity integration include advanced batteries, inverters, advanced controllers, and smart technologies [40].

In the case of clean water, drivers that support water security are water conservation, water recycling, and efficient water use. A number of mechanisms are available for conserving water. Typically, groundwater aquifers collect less than 40 million m^3 annually as natural recharge. This imbalance makes the need for changes and rethinking the water cycle obvious. Minor changes such as changing a showerhead to a more efficient one can save small amounts of water at the household level. Hence, the impact of such changes is limited if one household is considered. This impact can be significant if large communities and neighbor- hoods are the bases of the analysis. Other opportunities include industries and commercial sectors taking the initiative to recycle and reuse both gray and black water on-site. A collective support of this kind from residential areas, industries, and commercial sectors can significantly impact the strain on the main grid freshwater supply.

Water use patterns are critical to any water conservation solutions. For example, in the urban areas in Portugal, the residential sectors were observed to have the highest water demand when compared with the industrial, commercial, and insti- tutional sectors [41]. Reducing the domestic water consumption rises important benefits like the postponement of investments in the water supply system expan- sion and pump nanogrid upgrade. It also reduces peak and average effluent loading to the wastewater system [42]. A significant reduction on energy requirement is caused by a lower water demand in the household (e.g., for water heating). In addition, the water end-use sector of a distribution system (i.e., activities that use water in buildings and homes) has been found to be the highest energy intensive part in the urban water supply systems [43]. Such analysis, data, and information gathering can provide useful insight into practical water end-use efficiency programs that can be used by utility companies for the benefit of national economies.

Many studies have been conducted, for example, by Loh and Coghlan [44], Willis et al. [45], Beal et al. [46], Matos et al. [47], Cole and Stewart [48], and Omaghomi and Buchberger [49], to describe and characterize the types of water uses. These studies show that water end-use characteristics generally differ from place to place. Hence, it is important to analyze water end-use within local context in order to develop tools, mechanisms, and techniques for improving water end-use efficiencies. Studies by Willis et al. [50], Matos et al. [51], and Hunt and Rogers [52] demonstrated the relationships between consumer sociodemographic characteristics, end uses and consumer attitudes, and water end-use efficiencies. Willis et al. [53], Lee et al. [54], and Carragher et al. [55] reviewed the effect and the influence of the residential water use efficiency measures on water demand.

Improving the collection of gray water might significantly decrease the amount of clean water that is being used in landscaping, gardening, and toilet flushing at the household level. In Qatar, for example, gardening consumes around 5% of the total freshwater at household level. Albeit, gray water produced at houses is usually sent down the drain in sewage pipelines. Although the amount at one household may seem small, it is the collective actions of all residential areas that will affect the main grid strain on freshwater supply for a national economy. In addition, a lot of gray water is produced in other places such as mosques, air-conditioning units, shopping malls, and corporate buildings. By rethinking this practice, gray water collected for recycling from different places can be treated using simple processes to make this water suitable for gardening, landscaping, agriculture, construction works, and district cooling services. In spite of these potential reuses and recycling possibilities, gray water and black water in the case study villa is currently being channeled into a shared sewage system, which makes the gray water highly unusable. Although Qatar has a huge and a broad network system for collecting and treating the domestic wastewater, separation of the wastewater at the source can be more beneficial and more cost-effective in the long run than the central collection and treatment practices in the case study.

Due to the problems and challenges faced by the water sector, a number of water intelligence tools have been developed worldwide to alleviate global water issues. Information and communications technology (ICT) offers valuable chances to improve the efficiency and the productivity within the water sector, with the purpose of contributing to the sustainability of the resource. The increasing availability of more intelligent, ICT-enabled means to manage and protect the water resources of various national economies has led to the development of smart water management (SWM). The SWM approach promotes the sustainable consumption of water resources through coordinated water management, by integrating ICT products, solutions, and systems, targeted at maximizing the socioeconomic welfare of a society without compromising environment [56]. In Qatar, the potential use of ICT-enabled technologies has been initiated by Ooredoo, the telecommunication company in a pioneer project that aims to make Qatar's Lusail City a smart city.

The concept of smart water involves gradual convergence and integration of ICT solutions applied within the water domain. The smart water concept seeks to promote a sustainable, well-coordinated development and management of water resources by the integration of ICT products, tools, and solutions, thus providing the basis for a sustainable method to water management and consumption.

An alternative way for more efficient water management could be offered by an approach that is fully linked to the quality of the vision developed for the Water Business Information System [57]. The more advanced ICTs used in the water system, the smarter the water becomes. For example, a water system with smart meters and smart pumps and valves is smarter than a system with smart meters only.

The concept of smart water and the level of water smartness depend on the number and the advancement of ICTs successfully implemented in the system. The implementation of the smart water concept has enabled significant improvement in water distribution, has helped to enhance wastewater and storm water management, and has helped to decrease losses due to nonrevenue water. The advantages of applying the smart water concept include increasing water quality and reliability, decreasing water loss due to leakage, reducing operational costs, ensuring proper management of green systems, and improving customer control and choice. At the household level, these advantages increase water end-use efficiencies, while at the national level, they increase the efficiency of the water sector and hence play a significant role in conserving water and thus reducing the strain on the main grid water supply and distribution [56]. A number of countries and communities have embraced smart water technologies [58–60]. Data obtained from implementation of smart technologies can help utilities in discovering problems on the consumer end of the water system. Consumption rates of water and power in Qatar are relatively high. This puts a large strain on the utility company. While the utility company has successfully implemented a number of projects to conserve water and power along the supply and distribution network, relatively few projects have been done in Qatar to reduce water and power consumptions at the household level. Most of the successful attempts at the household level have been through plumbing fixtures and conservation programs aimed at making residence aware of the need to conserve both water and power through Tarsheed, a proponent of the local water and power company.

Methodology

Both qualitative and quantitative approaches were used to synthesize the smart water/power nanogrid for households. Data was collected from a case study villa in Qatar. Analysis of the current situation revealed that at the household level, a number of factors influence water and power consumptions in Qatar. One of the factors is the water and power technologies implemented at the household level.

Although a number of water and power saving tips have been provided by the utility company, no strict rules, regulations, or policies that directly influence water and power consumptions at the household level are available, although there are plans underway. Typically, most houses in Qatar are water and power metered, but the water and power rates are heavily subsidized by the government. In addition, no smart metering was available in households at the time when this project was carried out, although plans for smart metering were in place in the developing smart city of Lusail. Moreover, no information devices and except readings from conventional meters were available at the household level. Among the various types of residential unit villas were selected for case study since they composed the vast majority of residential preference of Qatar's residents. With various customer sectors, the focus on residential units also stems from its consumption contribution, with almost 90% of the national water consumption concentrated in residential areas [61]. A description of the case study villa is given in the following section.

Case study description

A typical three-storey villa was chosen as a case study in the development of the proposed smart water/power system. At the villa, consumption points include the bathroom and kitchen sinks, bathtubs, toilet, washing appliances, and garden watering. Wastewater from the houses is classified into two categories, i.e., black water (water produced from toilets and bidets) and gray water (water produced from all sinks, showers, and bathtubs). Black water and gray water are not separated in the current household water network outlet piping but are collected into one sewer pipe before disposal into the main sewage network. It is important to note that most villas in Qatar have a flat roof and there is no provision for harvesting rainfall since there is very little rainfall. In addition, vertical roof-mounted tanks are a common site in most villas in Qatar. Usually, a camouflage or protection structure is provided to make the roof tanks less visible. It is also important to note that typical families in Qatar are relatively large, with an average of 10 incumbents. A pictorial, schematic, and plan view of the case study villa is shown in **Figure 1**.

(a) Pictorial view   (b) Schematic side view

(c)

Figure 1. *Case study villa: (a) pictorial view; (b) schematic side view; and (c) schematic plan view of the main water and electricity consumption points in a typical villa.*

Assumptions

Based on survey results, a number of assumptions incorporated into the analysis were made as follows:

- A typical house in Qatar is designed for a family of 10 people.

- Consumption of water is the same in all households, as they are averaged per residential unit.

- All household pipes can be customized to meet the functional requirements of smart water and power technologies. This allows retrofitting.

- All smart water and power technologies will not cause any interference or degradation of the water and power quality or services provided by the main grid components.

Data collection

Sources of data include data logging of water and power consumption data, in-home interviews, as well as water and electricity meter billing data. Smart meters and a logger were installed in the case study to collect flow observations from each water consumption point in the house. The collected data was used to determine the various water consumption events in the household such as volume, average flow, and maximum flow. Water and power consumptions are the amounts of water and power that reach consumers or end users and are usually estimated by water or electricity meters at the consumer and end-user points.

Water and energy consumptions

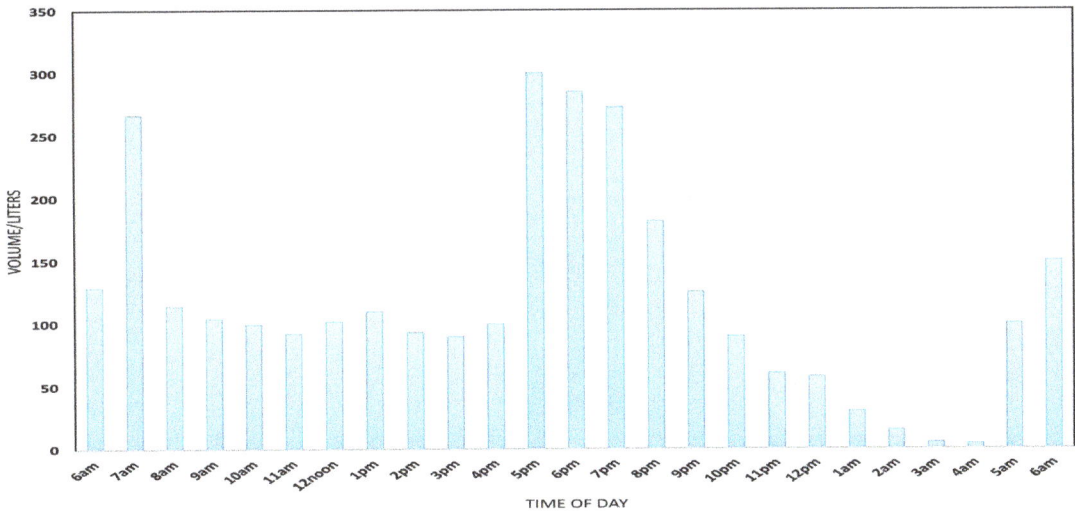

Figure 2. *Water consumption for a typical day at the case study villa.*

Pattern of water consumption on a typical day for the case study villa is shown in **Figure 2**. Typical monthly water consumption for the case study villa is also shown in **Figure 3**. **Figure 2** shows that water consumption per day varies from hour to hour depending on the needs of the people in the household. For example, **Figure 2** shows peaks at certain times (e.g., 7 am and 5 pm–7 pm) of the day corresponding to the times when water is required by most people in the household.

Figure 3 shows that the daily water consumption in the case study varies greatly from day to day. For example, it can be observed that there are peaks of water consumption at regular intervals throughout the month corresponding to high water consumptions. Interviews with residents in the case study villa revealed that more water is required on these respective days of the month for other uses such as various types of cleaning activities. Although variations are inevitable, the analysis in this work is based on the fact that there is a consistency in the flow patterns of residential water uses [62]. Pattern of power consumption on a typical day for the case study villa is shown in **Figure 4**. Typical monthly power consumption for the case study villa is also shown in **Figure 5**. **Figure 4** shows that power consumption per day varies from hour to hour depending on the needs of the residents. For example, **Figure 4** shows peaks at certain times (e.g., 7 am–9 am and 12 noon–8 pm) of the day corresponding to the times when power is required most in the household. **Figure 5** shows that power consumption per month varies from day to day depending on the needs of the residents. For example, **Figure 5** shows peaks on certain days of the month corresponding to days when power is required most in the household.

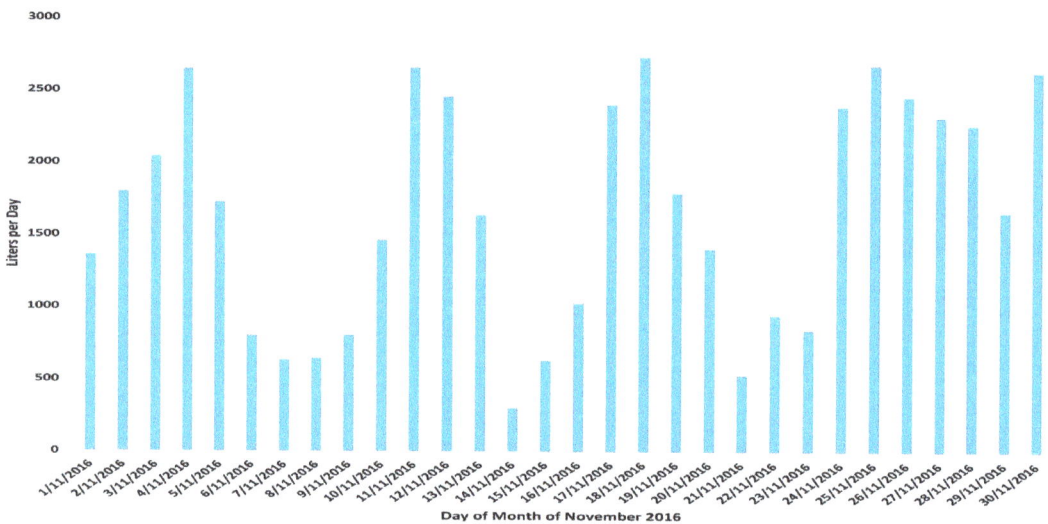

Figure 3. *Water consumption for a typical month at the case study villa.*

Water and power end-use fractions

It has been observed that the majority of Qatar's water consumption is centered on residential areas [35]. Hence, more has to be done to conserve water in residential areas. The main sources of leakages in households are the faucets, toilet seats, bidets, showerheads, tubs, and junction points between pipes. Some of the reasons cited for these leakages include different types of materials used in pipping, changes in different pipes sizes, high water pressure at junctions' points between pipes, and the materials' corrosions of pipes. Besides losses at these leakage points, a lot of water is used by residents for various reasons. **Figure 6** shows the daily water and power fraction end use for the case study villa.

From **Figure 6**, it can be observed that the main points of potable water consumptions in the house are bathing, personal washing, and toilets. Bathing contributes 43% of the total daily potable water consumed in a typical house, while air conditioning contributes 60% of power consumed in a typical house in Qatar. Since air conditioners are used most of the day during summer, a lot of condensate is drained and redirected into the sewer line as gray water. The proposed household nanogrid collects gray water from various consumption points and redirects it for reuse.

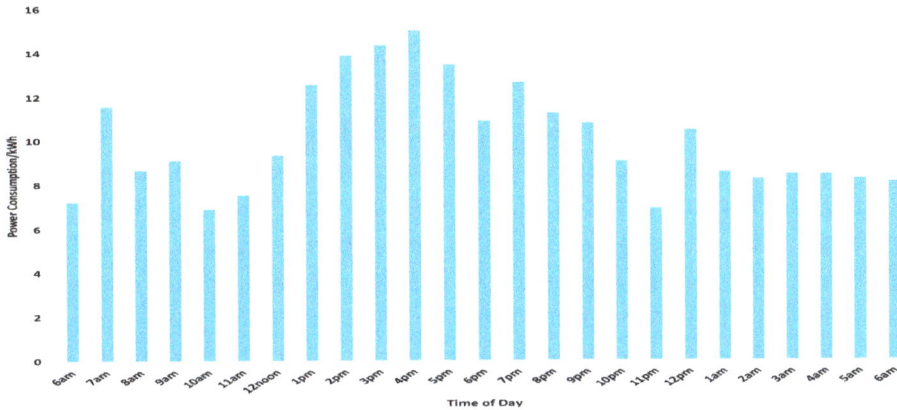

Figure 4. *Power consumption from the main grid for a typical day at the case study villa.*

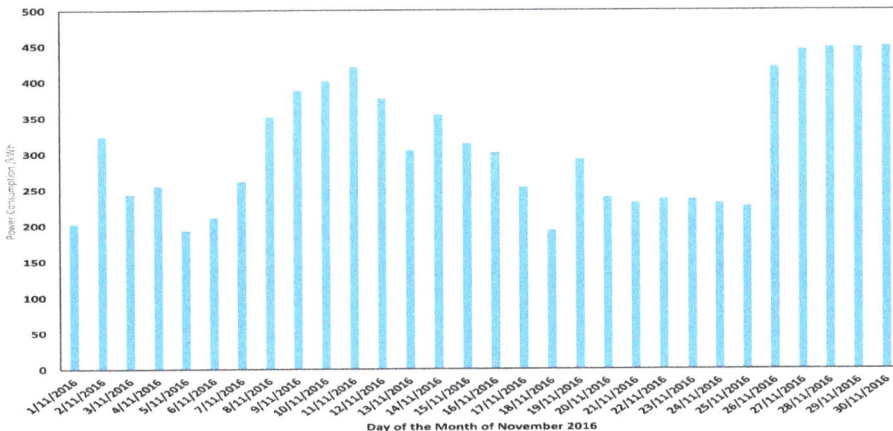

Figure 5. *Power consumption for a typical month at the case study villa.*

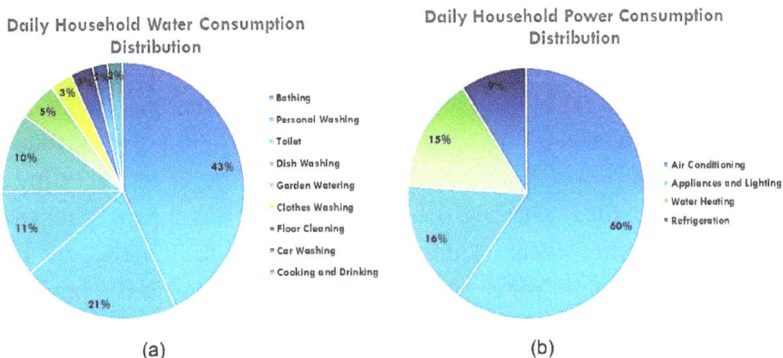

Figure 6. *Typical household freshwater and power use.*

Design and analysis

The design analysis presented in this section is based on the information obtained from the case study villa. Proposed design perspectives are based on rethinking the urban water and energy cycles. The theme is devised based on methods, techniques, and technologies for transforming the current water and energy infrastructures into a completely redesigned setup based on the following concepts: multi-utility loops, smart water and energy, integrated gray water infrastructures at the household level, and separation of water resources at the household level.

Multi-utility loops

Multi-utility design features can be used to collect data from all types of smart meter installations. This requires implementations of multi-utility metering and multi-utility controllers that ensures security in data communication. For the case study, the proposed design implies a system that enables multiple loops for multiple alternative water sources, i.e., water from the main utility grid, water collected from various consumption points in the house, and water collected from air-conditioning units. Water from the household consumption points and water from the air-conditioning systems are gray water that can be treated on-site. The aim in implementing the multi-loop system is to ensure maximum use of water available at a household. This requires additional piping as well as wastewater treatment systems. The objective is to design and implement a water nanogrid that minimizes the carbon footprint of water use and reduces water leakages. The electricity network is envisaged to have multiple loops, each loop representing the source of on-site energy generation. Available power loops that have been included in the analysis are main power grid, solar PV, and in-pipe hydroelectricity. In such a design, a multi-utility controller will enable communication among the various consumption meters installed in the house in order to determine how much water or power is at disposal. Several costs are involved when upgrading the current water and energy infrastructures in households. Such costs include cost of smart meters, cost of installation and maintenance, as well as costs of data communication tools.

Design requirements

The dual-purpose smart water/power nanogrid is envisaged to be made up of water and energy smart technologies integrated into a nanogrid for household use. The purpose of such a nanogrid is to help in conserving water and energy. Such a nanogrid includes on-site power generation, on-site water recycling and reuse, as well as communication interfaces that will provide real-time information to house- hold users about water and energy consumption levels. On-site power generation will reduce dependence on the main power grid and hence alleviate the strain on balancing power supply and demand. On-site water recycling will reduce consumption of freshwater, thus relieving the strain on freshwater supply networks. Information provided through the communication interface is expected to influence the behavior of house-hold users in terms of sensible water and energy consumptions at the household level. Design parameters were collected from the case study. Based on the results of a survey carried out in a residential community in Qatar as well as the survey from the utility company, a number of design requirements for a smart water/ power system at the household level were identified and are summarized as follows:

1. Minimize water and power consumptions at the household level in order to (a) reduce water and electricity bills for household owners and (b) reduce the strain on the main power and grid for utility company

2. Implementation of renewable energy technologies to alleviate the strains on the main power grid

3. Minimum breakdown of the system in order to ensure reliability in water and energy supply

4. Remote control for managing the water/power system with respect to house- hold owner choice and preferences

5. Real-time notification about water and power consumption status for house- hold owner decision-making

6. Ease of retrofitting the multi-loop nanogrid

7. Having an "acceptable" cost of procurement and installation of the system

8. Minimum safety threats to household users of the system

The conceptual design of the smart water/power system consists of different types of components. System design parameters for visualizing the architecture of the proposed smart water/ power system were derived from the general nanogrid concept, i.e., nanogrids are autonomous renewable energy systems that do not interfere with the main grid. This consideration was important since currently the utility company in Qatar does not allow transfer or sharing of power across the main grid. The conceptual extension of nanogrids relates to an integral nanogrid composed of both smart water and energy technologies. The combined inclusion of smart water and power technologies in one nanogrid constitutes an important nanogrid design worth pursuing. With such nanogrids in place, it will be easier to translate existing nanogrid into a functional smart microgrid. Technologies selected for the nanogrid include solar PV, reverse osmosis, pumped storage, inpipe hydropower generation, as well as energy storage components such as batteries. Target design specifications for the smart water/power system were derived as follows:

1. On-site generated power must be able to supply at least 20% of total household energy requirements (based on the Qatar National Vision 2030 aspirations).

2. Solar PV panels (monocrystalline type—as per utility company preference) are to be roof mounted, and the total area of these panels must not exceed the roof area ≈ 400 m² for a typical villa in Qatar.

3. Since the AC load will be provided by the main power grid, there must be two controllers in the system: one for the DC load and another one for energy storage.

4. In-pipe generator unit has to be installed within 25 mm of water supply pipes in the other floors of the house based on the current water reticulation pipe network system in the case study.

5. In-series generator units must have a distance of 4× pipe diameter apart.

6. Reverse osmosis unit should be able to process gray water with the following parameters: temperature ≈25°C and pressure ≈1.5–7 bar. The treated gray water must be suitable for flushing toilets, gardening, and on-site landscaping.

7. The system must include the following components to enable a certain level of smartness: smart meters, smart valves, pumps, and pH sensors.

8. Selected system components must be able to communicate with components in the existing household nanogrid.

9. The water/power system must be able to send information to household users for decision-making.

10. System components should be able to work based on the following specifications: 240 V, electrical frequency of about 50 Hz, and 900 GSM frequency (as per Qatar specifications).

The smart water/power system consists of three main units: (i) on-site power generation, (ii) on-site gray water recycling and reuse, and (iii) communication unit that will provide users with information about water and energy consumption as well as quality of the recycled gray water. Technologies used to assemble the smart water/power system include in-pipe electricity generators, pumped storage, solar photovoltaics, reverse osmosis, and a control system. The in-pipe electricity generator will be used to produce electricity by utilizing the water pressure as the water moves through the water supply pipe network as well as the gravity from the pumped storage. The roof-mounted tanks will facilitate pumped storage that will be used to maximize the use and reuse of recirculated water in the household. Solar photovoltaic panels will be used to generate solar electricity. In cases when the power generated on-site is not sufficient, the main grid power will be used instead.

A reverse osmosis unit was used to treat gray water to sufficient quality for use in watering gardens, landscaping, car washing, or flushing the toilets. A control system was used for managing the operation of components and devices in the system as well as to provide household users the information on the system status. Smart meters were used to digitally send meter readings to household users so that they know their water and power consumptions will be added. Smart shut-off valves were used to facilitate remote control of water in the household. pH sensors were incorporated to facilitate the effective control and communication of the water quality in the system. A plumbing network, additional to the existing infra- structure, was used for water circulation in the system. **Figure 7** shows a schematic representation of the proposed system components.

Most houses in Qatar have little space available outside the building area. Therefore, the best place for the solar panels and storage tanks is on the roof. For practical implementation in the case study, these components must meet the minimum standards stipulated by the local utility company. The standards for the pipe network type and materials are controlled by the available local construction standards, codes, and regulations. As further requirement, additional components of the communication network must not degrade performance of currently existing infrastructure.

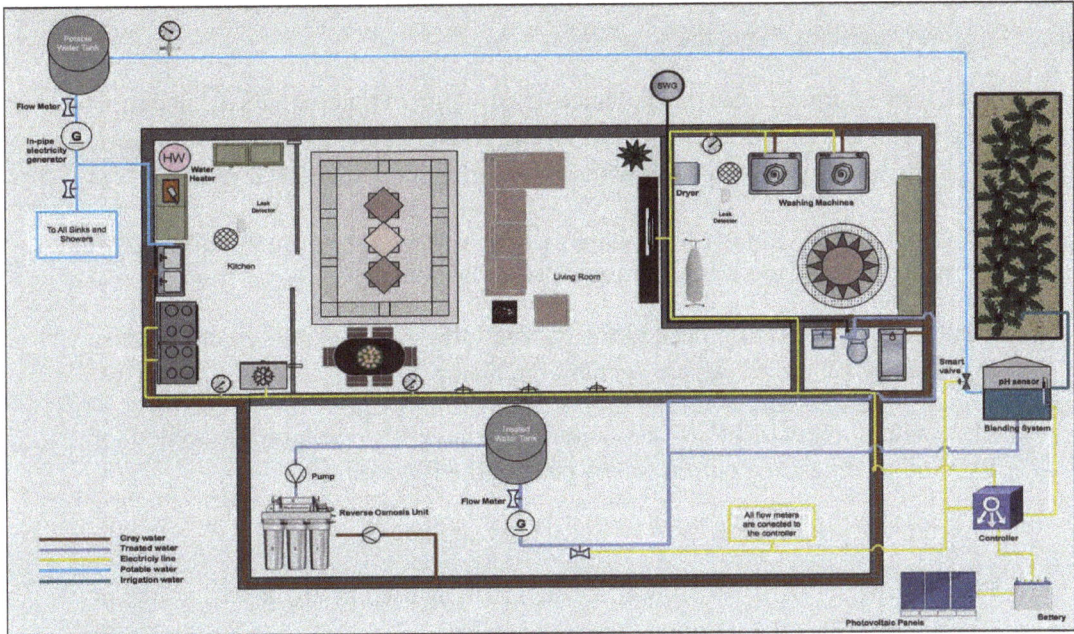

Figure 7. *Schematic diagram showing position of the various technologies implemented in the project.*

On-site power generation

The major components for on-site power generation are a solar PV system and in-pipe hydropower generation. The solar PV components include an array of PV modules, a charge controller, inverter, and battery bank. In pipe hydropower, generation was considered for both the existing network and the auxiliary network meant for recycling gray water at the villa. Available in the existing network is a pump that pumps water to the roof tank. The movement, flow, and velocity of this pumped water are captured to form the first type of in-pipe hydropower generated on-site. When the gray water is recycled, through the reverse osmosis unit, the water is pumped to the rooftop so that it can be conveniently used by taking advantage of the gained potential energy.

Design of system elements

Solar PV system sizing

In order to size the solar PV system, the following steps were followed:

1. Calculate the total power for all loads that will use solar PV electricity by adding the total watt-hours for each appliance used and finding the total watthours per day needed from the PV modules.

2. Sizing the PV modules by calculating the total watt-peak rating needed for PV modules and finding the number of required solar PV panels.

3. Sizing the inverter.

4. Sizing the battery.

Eqs. (1)–(6) were used for sizing the solar PV system.

Total watt – hours per day for appliances used = Sum of watt – hours for all household appliances (excludes microwave ovens, cooker and any electrical machines, AC units, and any e lectric tools with power rating more than 1500 W)   (1)

Total watt – hours per day provided by PV modules = 1.3 × sum of watt – hours for all appliances (to cater for energy lost in the system)   (2)

Total watt – peak rating needed for PV modules = total watt – hours per day provided by PV modules/panel generation factor   (3)

Minimum number of PV panels required = total watt – peak rating needed for PV modules/the rated output watt – peak of the PV modules   (4)

Inverter size = 1.3 × total watts of household appliances   (5)

Battery capacity (Ah) = total watt – hours per day used by appliances × days of autonomy (0.85 × 0.6 × nominal battery voltage)   (6)

The requirements for the solar PV system and parameters used for sizing the solar PV system are shown in **Table 1**.

Table 1. *Solar PV requirements and design parameters.*

Parameter	Value	Solar PV system components	Specifications
Total watts for appliances	1500 W	**Household power consumption demands**	55 kWh
Total watt-hours for lighting and appliances per day (excludes microwave ovens, cooker and any electrical machines, AC units, and any electric tools with power rating more than 1500 W)	55,000 Wh	Total watt-hours per day for appliances used	71.5 kWh
		Total watt-hours per day to be provided by PV modules for house appliances	
Panel generation factor	5.84		
Days of autonomy	2	**PV modules size**	7.15kWh
Nominal battery voltage	12 V	Total watt-peak rating needed for PV modules	200
Nominal panel wattage	350 W	Minimum number of PV panels required	
Battery loss	0.85	**Size of inverter**	1800 W
Depth of discharge	0.6	**Size of battery**	11,000 Ah

Pico hydroelectricity

In-pipe hydropower (or pico hydroelectricity) represents a clean source of energy that focusses on recovering the energy used to supply water to house- holds. The energy used to treat gray water can also be partially recovered by taking advantage of pumped storage. In pico hydroelectricity generation, a turbine is forced to rotate due to flow and pressure of water in a water pipe network. The rotating turbine is connected to a generator that generates electricity. This technology has been successfully implemented in various contexts [62–64]. The amount of power generated at the household level is relatively small [65–67]. However, this amount becomes significant to the utility provider if the technology is implemented in all houses as a national level project. Since in-pipe generators are preferred in the aboveground location with gravity-fed delivery pipelines, their position outside the villa's walls is ideal for maintenance and requires minimal changes to the system's operations when retrofitted. The in-pipe generators were designated to the main supply pipe based on the following criteria:

- Size: The main supply pipe is installed with the largest diameter among the gravity-fed plumbing pipes, thus allowing the installation of the largest possible generator to yield the highest possible power.

- Pressure: The pressure within the supply pipe is maintained by the roof pump at 20 psi, with a maximum pressure head at the ground floor before it reaches the first floor's outlet.

- Water quality: The water flowing through the main supply pipe is potable and does not carry the risk of containing debris, solid waste, or mixed fluids (e.g., oil from the kitchen) that may affect the operations of the turbine blades or cause damage to its physical structure.

Assumptions:

- The in-pipe generator's installation begins 0.25m above the ground level to avoid the interference with the distribution inlet of the first floor.

- The in-pipe generators operate with an efficiency of 65%, a practical estimation since in-pipe generation is a proven technology with high reliability and capacity, allowing it to maintain high efficiency even when facing variable flows.

- Cumulative water consumption occurs in the house for 8 hours a day.

The number of in-pipe generators to be installed is determined with consideration to:

- Spacing factor: In the current installation procedures, it is recommended by most suppliers to space the generators by a 4- diameter factor. The spacing is to prevent generator from affecting the functionality of the following one; such effect includes allowing the water turbulence to dissipate.

- The loss of power generated: When positioning the generators 4 diameters apart, the height of the water column causing the pressure head will decrease, resulting in a drop in the power produced between successive in-pipe generators.

Eqs. (7)–(10) were used to calculate hydropower generated at the household level. **Table 2** shows the design parameters for the hydropower generation.

Total wattReynolds number, Re= $(Q \times D)/(v \times A)$ $(0.0017 \times 0.035)/(1.004 \times 10^{-6}) \times (9.62 \times 10^{-4}))$ $\qquad$ (7)

Head loss across pipe, $hl = (16/Re) \times (L/D) \times (V^2/(2 \times g))$ $\qquad$ (8)

Head loss due to turbine, $ht = (z) - (V^2/(2 \times g)) - hl$ $\qquad$ (9)

Power output per in $-$ pipe generator $= Q \times ht \times W \times \eta$ $\qquad$ (10)

A typical multi-loop power network for the case study is shown in **Figure 8(a)**. **Figure 8(a)** shows multiple power flows from three different sources: main power grid, solar PV, and in-pipe hydropower. **Figure 8** also shows converters that facilitate the use of generated power in the household depending on whether the appliance requires AC or DC power. A typical multi-loop water network for the case study is shown in **Figure 8(b)**. **Figure 8(b)** shows multiple water flows from two different sources: main water supply from utility and flow of recycled gray water.

Gray water recycling

Table 2. *Design parameters for the hydropower generation.*

Parameter	Value
Flow rate of the water (Q)	$0.0017 \, \text{m}^3/\text{s}$
Velocity of the water (V)	$1.84 \, \text{m/s}$
Diameter of the pipe (D)	$0.35 \, \text{m}$
Area of diameter (A)	$9.62 \times 10^{-4} \, \text{m}^2$
Total length (L)	$11.9 \, \text{m}$
Dynamic viscosity (μ)	$1.002 \times 10^{-3} \, \text{N s/m}^2$
Kinematic viscosity (v)	$1.004 \times 10^{-6} \, \text{m}^2/\text{s}$
Specific weight of water (W)	$9790 \, \text{N/m}^3$
Efficiency of in-pipe turbine (η)	0.65

Gray water recycling was achieved by installing a reverse osmosis (RO) unit. Requirements for the feed water include the water pressure inside the pipes, the quantity of gray water to be treated, and the temperature of feed water. The size and quantity of membranes required to produce the desired volume of permeate were selected based on off-the-shelf units. **Table 3** shows a comparison of the properties of feed water data at the household level, reverse osmosis requirements, and local authority requirements for recycled gray water. From **Table 3**, it can be observed that the feed water temperature is 28°C, i.e., 3° more than the required. Although this difference will have an effect on the quantity of treated water, the produced quantity is expected to be within the range of that stipulated by the local authority. The range of pressure (95–100 psi) is suitable for the reverse osmosis unit since it is high enough to allow all the solutes to be rejected from the solvent, thus creating treated gray water with the required specifications.

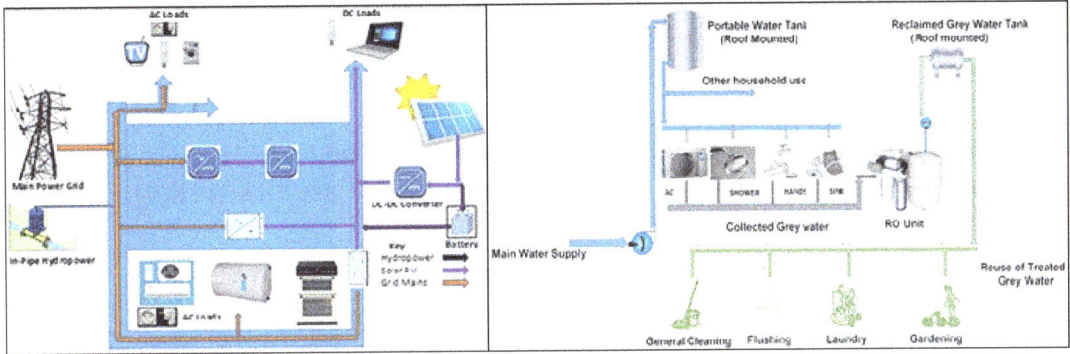

Figure 8. *(a) Multisource power loops and (b) multisource water loops for a typical household with solar energy and nano-hydropower.*

Table 3. *Comparison of feed water data with reverse osmosis and local authority requirements.*

Constituent	Gray water	Reverse osmosis requirements	Requirements of local authority
Total solids	700 PPM	<1000 PPM	≤5 PPM
Chloride	50 PPM	<0.1 PPM	≥1 exiting treatment system, ≥0.2 at user end
Alkalinity (as CaCO₃)	100 PPM	<50 PPM	N/A
Biochemical oxygen demand (BOD5)	200 PPM	N/A	≤10 PPM
Turbidity	2–5 NTU	<1 NTU	≤5 NTU
Temperature	28°C	25°C	N/A
Pressure	95–100 psi	21.8–101.526 psi	N/A
pH	7.5	3–11	6–9
E. coli	0	0	Non-detectable
Dissolved oxygen in reclaimed water	N/A	N/A	≥2 PPM
Color	N/A	N/A	≤20 Hazen unit
Threshold odor number (TON)	N/A	N/A	≤100
Ammoniacal nitrogen	N/A	N/A	≤1 mg/l as N
Synthetic detergents	N/A	N/A	≤5 mg/l

Since the total dissolved solids (TDS) of gray water is less than the TDS of what the feed water supposed to be, no filter was required for reducing the total dissolved solids. However, a sediment filter was required in order to remove dust, sand, suspended solids, particles, and rust, down to 5 μm. The hardness of the feed water is higher than that required. Hence, a hardness filter (water softener) was required in order to decrease the hardness of the gray water so that the reverse osmosis unit can function and produce treated gray water with the desired hardness limits. The concentration of chlorine in feed water is too high. Therefore, an activated carbon cartridge prefilter was required in order to minimize the level of chlorine in the gray water to an acceptable range before it goes to the granular activated carbon filter. The granular activated carbon filter was used to get rid of unpleasant chlorine, tastes, odors, cloudiness, colors, organic chemicals, sulfur, suspended particles, and dirt. This filter will also reduce the amount of the chlorine in the water to a desired value. Since the turbidity of feed water is high, a micro cartridge filter was used in order to reduce the turbidity of the gray water. This will help in achieving the required turbidity in the treated gray water.

Communication unit

Smart meter sensors allow water consumers to gain information on the water usages, on the water leaks, and on the quantity of water that is being drawn from the main grid and consumed in the house. This information is expected to allow users to control water leaks and abnormal usages. The implemented sensor is noncontact with water and makes use of the "pulse output" facility that is built in to most water meters. The smart meter sends information related to the water flow and the water quantity withdrawn from the grid to the control unit using a wireless connection.

The control unit sends this information to the user's phone at regular predeter- mined intervals to indicate real-time water consumption from the water meter as well as provide visual and sound alerts if there is an incident such as an abnormal water usage. Water leak detection is programmed to notify the users of the house when there is a water leak at a specific point in the house by producing a sound alert and an SMS for the user to take an action. If the user does not take any action after 10 min, a signal will be sent to the control unit via a wireless connection. In return, the control unit will send a message to the user's phone asking him/her to take an action as a soon as possible. The smart valves are smart due to their ability to open and close automatically based on specified conditions and commands from the control unit. As an example, pH sensors are used to measure the pH of the treated gray water. The pH sensor plays an important role in blending the treated gray water. Water blending was required to ensure that the pH of the treated gray water is suitable for garden watering and other applications. If the pH of the treated gray water is less or more than the required value, the pH sensor sends a signal to the control unit to open the smart valve in order to allow potable water to flow from the potable water tank to treated gray water tank. This flow is expected to neutralize the pH of the treated gray water so that the gray water is suitable for different purposes.

Construction and testing of the prototype
Prototype materials and assembly

In order to realize the functionality of the proposed design, a prototype was constructed. Components for the smart water/power system were assembled from standard components available off the shelf. The prototype construction included the physical structure and the control system. The physical structure consists of the positioning of devices such as in-pipe generators and flowmeters in addition to the plumbing network. The control system was assembled from Arduino Mega boards, and the control actions were programmed using the C

language and the Arduino software. Prototype components included pumps, wastewater tank, treated water tank, a battery, storage tanks, reverse osmosis unit, in-pipe electricity generator, flowmeters, smart valves, photovoltaic panels, pipes, flexible hose, pipe fittings, sensors, pH meter, Arduino Mega board, and a GSM shield board. The selected pump has a voltage of 12 V, so it can be connected to a battery of 12 V, since this 12 V battery will supply the prototype with electricity. The battery was continuously charged by a 100 W solar panel. The suction lift of this pump is 1.2 m, which means that this pump will be able to pump the water to the storage tank at a height of 0.91 m and circulate water in the pipe net- work for simulated water use in the house. The voltage of the in-pipe electricity generator is 5 V, which is compatible with the battery and to the Arduino, which can take a maximum of 5 V. The in-pipe electricity generator has a water pressure of 0.05 MPa, which is suitable for the pipe dimensions used to construct the prototype. Reverse osmosis unit was chosen to have specifications that depend on the flow rate and the pressure of the water within the pipe networks. The reverse osmosis unit used in the prototype has a maximum capacity of 280 L/day, which is sufficient for prototype demonstrations. In addition to measuring the flow of water, flowmeters were used as devices to detect the leakages within the pipes.

This was done by installing two flowmeters at a junction point or "leak hole" to simulate water leakages. Differences in the flowmeter readings would indicate a leakage. The function of smart shut-off valve was linked to that of flowmeters in such a way that when there is a difference between the values of flowmeters at leak points the smart shut-off valve will stop water flow in the pipe. **Figure 9** shows a pictorial view of the assembled prototype as well as a sample of main prototype elements.

Experiments with in-pipe hydropower demonstrated that electricity was produced and used to light a bulb in the prototype. Initial testing of the water section of the prototype included running potable water through the prototype with normal flows. It was observed that the reverse osmosis unit was taking too long to treat gray water, due to low pressure. The low efficiency and slow speed processing of the reverse osmosis unit was identified as a bottleneck in processing gray water. Gray water was supplied to the reverse osmosis unit to determine if the quality of the treated gray water was good enough for its intended purposes. Treated gray water parameters were found to be 4.3 ppm for total dissolved solids, 10 ppm biochemical oxygen demand, and a pH of 7. These values are close to the treated gray water requirements as stipulated by the local authority. Further testing of the prototype was done with the simulated pipe leak and running both portable and gray water in the prototype. The results of this test showed that the proposed method for identifying water leaks was suitable since differences in flowmeter reading were observed when a simulated leak occurs. The prototype was also able to send user notifications (to a smart phone) regarding the condition of the treated gray water and the presence of a leak.

Case study results

After experimenting with the prototype, the main components of the nanogrid were installed parallel to the existing infrastructure at the case study villa. The parallel installation was designed to replicate functionality of the prototype as well as for minimum interruption of normal household activities as per the requirements of the household owner. In addition, the parallel installation allowed easy removal of installed components after data collection. The following subsections summarize the analysis of data collected from the case study.

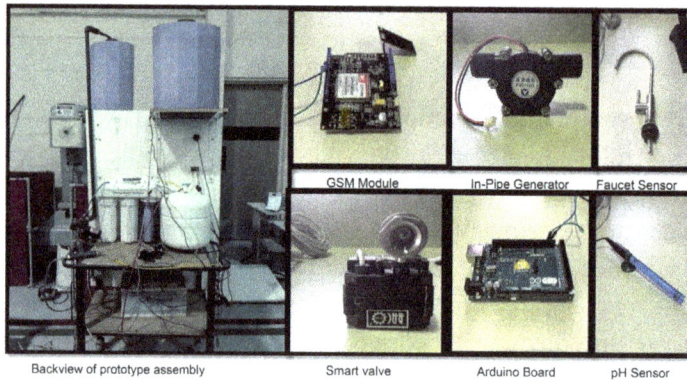

Figure 9. *Assembled prototype and a sample of the main components.*

In-pipe hydropower generation

Table 4 shows the expected power to be generated from the in-pipe hydropower generators installed on the main water supply line.

Table 4. *Expected power output from case study villa.*

In-pipe generator	Elevation (mm)	Distance from the roof (m)	Losses across pipe (m)	Turbine losses (m)	Power output (kW)	Estimated daily output (kWh)
1	25	11.90	0.015	11.70	0.0819	0.655
2	165	11.76	0.015	11.60	0.0813	0.650
3	305	11.62	0.014	11.40	0.0806	0.645
4	445	11.48	0.014	11.29	0.0793	0.634
Total					0.3231	2.585

From **Table 4** the insertion of four in-pipe generators provides the house with kWh of electricity, for 8 h of water consumption per day. Taking in account assumptions and constraints, four in-pipe generators were positioned as follows:

- Supply in-pipe generator 1 at elevation 25 mm, at the lowest point considering the assumed 25 mm clearance.

- Supply in-pipe generator 2 at elevation 165 mm, 140 mm apart from generator 1 after spacing an amount of 4 diameters.

- Supply in-pipe generator 3 at elevation 305 mm, 140 mm apart from generator 2 after spacing an amount of 4 diameters.

- Supply in-pipe generator 4 at elevation 445 mm, 140 mm apart from generator 1 after spacing an amount of 4 diameters.

Since the water tank will be placed on the roof, power and placement of the inpipe generators for the recycled gray water will be a replica of that shown in **Table 4**.

The total power generated from the recycled gray water's main water supply pipe is 2.59 kWh. Therefore, the total potential power generated from in-pipe hydropower generators at the case study villa is 5.17kWh (i.e., 155.06 kWh per month). These values agree with other research findings [68–72].

On-site solar PV power generation

The specifications for the solar PV power generated on-site are shown in **Table 5**. From **Table 5**, solar PV panels cover 40% of the total roof area. This leaves enough space for pumped hydro tank, AC units, and other equipment. From **Figure 6**, 60% of the energy at the household level is used for air conditioning, and this will be provided from the main grid power.

Table 5. *On-site power generation.*

System component	Quantity	Comment
Number of panels	200	At 350 W
Performance ratio	0.62	Experimentally estimated loses {AC/DC conversion loses, shading effect, dust, temperature effects}
Efficiency of panels	20%	Manufacturer specification
Power output per month from PV system	2145 kWh	Since the basic appliances require 1650 kWh per month, the solar PV system is more than able to meet the power requirements for these appliances
Total power required for the case study villa per month	9251.48 kWh	Estimated from case study data (see **Figure 5**)
Power from pico hydro generation	155.06 kWh	Includes generation from the main water supply and treats gray water circulation at a household
Total on-site generated power per month	2300.06 kWh	Pico hydro + solar PV
Percentage (%) contribution of power generated from solar PV	22.86%	This percentage (%) is estimated for 1-month operation. Due to the intermittent nature of solar energy and its dependency on climate, on the long run, this percentage (%) is expected to decrease
Percentage (%) contribution of power generated from pico hydro	2%	
Percentage (%) contribution of power generated on-site	24.86%	This percentage (%) is estimated for 1-month operation. Due to the intermittent nature of solar energy and its dependency on climate, on the long run, this percentage (%) is expected to decrease

Water savings

Water saving calculations per month for the case study are shown in **Table 6**. From **Table 6**, total freshwater savings amounts to 25% per month due to reuse of on-site treated gray water. This calculation takes into account losses in gray water collection system, efficiency of equipment used, as well as water loses during treating and recycling of gray water on-site. Efficiency improvements in the gray water collection network and gray water treatment systems can increase the total amount of reusable gray water with additional benefits at the house- hold level.

Table 6. *Water saving calculations based on case study villa data.*

	Quantity (liters per month)	Gray water generated (liters per month)	Recycled grey water (liters per month)	Freshwater savings (liters per month)
Total freshwater use in the case study	49755.49			
Toilet flushing	5473.12			
Gardening	2487.77			
Floor cleaning	1492.66			
Car washing	995.11			
Cooking and drinking	995.11			
Total unrecoverable	11443.77			
Bathing	21394.86			
Personal washing (hands and face on sinks)	10448.65	6791.62		
Dish washing	4975.55	3234.11		
Clothe washing	1492.66	970.23		
Total air-conditioning requirements for 1200 m2 (three-storey villa at 400 m² per floor area) = 361.34 kW. For 3.5 kW, 0.125 liters per kWh of condensate is collected from the case study villa = 45.17 l Total condensate collected per day (considering 8 working hours) = 361.36 l AC condensate	10840.8			
		6504.48		
Total gray water generated (losses in gray water collection system)	17500.44			
Total recycled gray water (loses during treating and recycling)			9975.25	
Total monthly water savings in case study				9975.25
Percentage (%) freshwater savings				20%

Economic analysis

The economic analysis was done for the case study villa. The proposed nanogrid at the household level is viewed as the responsibility of the house owner. Therefore, investment costs are borne by the household owner for individual villas. While economic benefits are important to the household owner, the utility company is also interested in how the proposed nanogrid can be used to mitigate the effects of climate change through a reduction of the energy and water consumption. The following equation was used in the cost analysis.

Simple Payback Period = Investment/Annual Savings　　　　　　　　　　(11)

Discounted Payback Period = $\{ -\ln(1-(\text{investment amount} \times \text{discount rate})/(\text{annual savings})\}/\{\ln(1 + \text{rate})\}$　　　　　　(12)

Monthly water bill as per local utility company tariff = (20) 4.4 + (49.76–20) 5.4 = QR 248.7

Monthly water bill savings = QR 248.7 – QR {(9.975) 4.4} = QR 204.81.

Monthly power bill as per local utility company tariff = QR (2000)0.08 + QR (4000–2000)0.09 + QR (6000–4000)0.1 QR+(9251.48–6000)0.12 QR = QR 930.18.

Monthly power bill savings = QR 930.18 – {(2000) 0.08 + QR (4000– 2000)0.09 + QR (6000–4000) 1+ QR (6951.42–6000)0.12 = QR 276.01.

Total monthly saving (water and power bill) = QR 480.82.

Table 7 shows the cost saving parameters in Qatari Riyals for retrofitting system components to enable multi-utility loops at the household level.

Table 7. *Cost saving parameters.*

Water	Annual cost saving QR
Total cost for the multisource water circulation system and recycling units (cost of smart water system components + pipe network + pumped storage tank + RO unit and accessories)	19,661.76
Annual water savings due to reuse of recycled water and smart metering	2457.72
Power	
Total cost of the multisource power loop system (cost of smart meter, solar PV modules, inverters, pico hydro components, and battery bank)	46,369.68
Annual power savings due to on-site power generation	3312.12
Total annual savings (water and power bills)	5769.84

The payback periods at a discounting rate of 10% from the case study household owner's perspective are shown in **Tables 8** and **9**. From **Tables 8** and **9**, it can be inferred that the payback period based on the simple payback calculation for smart water metering (7.25 years) is less than that of smart energy metering (9.59 years). As expected, the value of the discounted payback period is always higher than that from simple payback period. Further improvements to these investment paybacks can be realized by improving the efficiency of the gray water collection and treatment system.

Table 8. *Payback period for smart water metering at the household level.*

Payback period		7.25 years	
Discounted payback period		12.71 years	
Cash flow return rate		6.79% per year	
	Cash flow	Net cash flow	Discounted cash flow	Net discounted cash flow
---	---	---	---	---
Year 0	QR-19,661.76	QR-19,661.76	QR-19,661.76	QR-19,661.76
Year 1	QR2,457.72	QR-17,204.04	QR2,234.29	QR-17,427.47
Year 2	QR2,531.45	QR-14,672.59	QR2,092.11	QR-15,335.36
Year 3	QR2,607.40	QR-12,065.19	QR1,958.97	QR-13,376.39
Year 4	QR2,685.62	QR-9379.58	QR1,834.31	QR-11,542.07
Year 5	QR2,766.19	QR-6613.39	QR1,717.58	QR-9824.49
Year 6	QR2,849.17	QR-3764.22	QR1,608.28	QR-8216.21
Year 7	QR2,934.65	QR-829.57	QR1,505.94	QR-6710.27
Year 8	QR3,022.69	QR2,193.11	QR1,410.11	QR-5300.16
Year 9	QR3,113.37	QR5,306.48	QR1,320.37	QR-3979.79
Year 10	QR3,206.77	QR8,513.25	QR1,236.35	QR-2743.45

Table 9. *Payback period for smart energy metering at the household level.*

Payback period		9.59 years	
Discounted payback period		23.18 years	
Cash flow return rate		0.76% per year	
	Cash flow	Net cash flow	Discounted cash flow	Net discounted cash flow
---	---	---	---	---
Year 0	QR-36,369.68	QR-36,369.68	QR-36,369.68	QR-36,369.68
Year 1	QR3,312.12	QR-33,057.56	QR3,011.02	QR-33,358.66
Year 2	QR3,411.48	QR-29,646.08	QR2,819.41	QR-30,539.25
Year 3	QR3,513.83	QR-26,132.25	QR2,639.99	QR-27,899.26
Year 4	QR3,619.24	QR-22,513.01	QR2,471.99	QR-25,427.27
Year 5	QR3,727.82	QR-18,785.19	QR2,314.68	QR-23,112.59
Year 6	QR3,839.65	QR-14,945.53	QR2,167.39	QR-20,945.20
Year 7	QR3,954.84	QR-10,990.69	QR2,029.46	QR-18,915.74
Year 8	QR4,073.49	QR-6917.20	QR1,900.31	QR-17,015.43
Year 9	QR4,195.69	QR-2721.50	QR1,779.38	QR-15,236.05
Year 10	QR4,321.57	QR1,600.06	QR1,666.15	QR-13,569.89

Toward a smart water/power microgrid

The smart water/power concept discussed in this chapter is at the household level. In the proposed implementations, smart water/power nanogrid is the smallest unit of a dual-purpose smart water/power distribution network that is capable of independent operation to support the main grid water and power distribution and utilization at the household level. This essentially represents a smart water/ power nanogrid composed of local small-scale generators of water (recycled gray water) and electricity (solar PV and in-pipe hydropower electricity). The proposed smart water/power nanogrid can be used to conserve water end use at the household level, thus relieving the strain on the main water grid as well as to supplement power supply at the household level, thus relieving the strain on the main power grid.

A network of smart nanogrids could be interconnected into a microgrid without any central entity. Thus, the proposed smart water/power nanogrid for a single household can be connected to another nanogrid of a neighboring house. A group of interconnected nanogrids can be configured into a microgrid, and a group of microgrids can be configured into a mesogrid as shown in **Figure 10**.

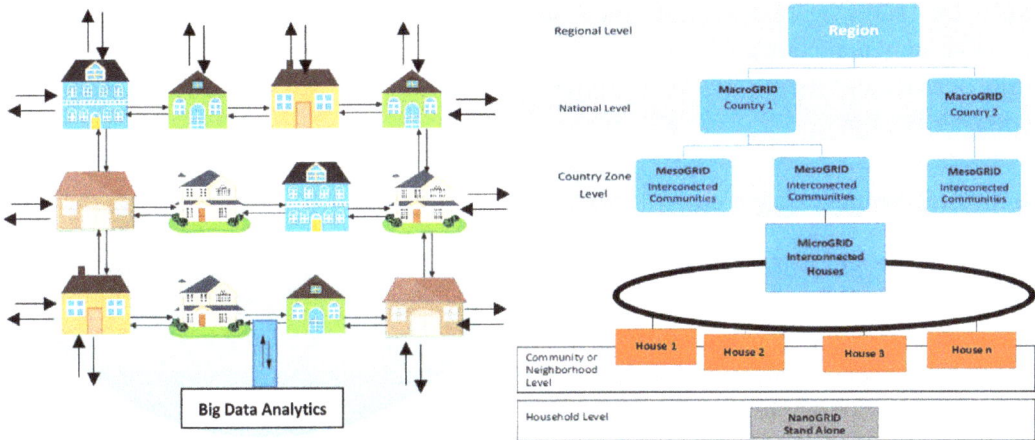

Figure 10. *Interconnected households.*

When extending the proposed nanogrid concept to microgrid and mesogrid, each smart residential unit is viewed as a single node with interconnectivity. Such interconnectivity provides the household users with water/power availability and consumption status in neighboring units through real-time user notifications. The information would include excess power and water generation in a neighboring node and neighbor's willingness to share such excess, along with the sharing conditions. The ability of the system to share provides the unit owner with the option of setting usage priority to or from the main grid or the mesogrid as preferred when the unit's power and water generation do not satisfy the user's demands. Such access promotes the status of the residential unit to that of a prosumer, a producer, and a consumer simultaneously. In the evolution process of the nanogrid toward the smart grid, the scales of water and power production are expected to increase. For example, simple water recycling and reuse at nanogrid can be expanded to fresh- water production sources (small-scale water desalination, bigger water treatment structures, and water reservoirs) supported by zero water discharge policies. Power production scales can evolve from the rooftop solar PV panels to solar PV arrays.

The increase in scales will help in stabilizing the water/power decentralization plans. A node's sharing conditions are dependent on the individual prosumer's discretion in terms of selling cost, quantity, and the threshold of personal consumption. The exchange of information between various nodes in the microgrid or mesogrid and the level of access is to be governed through applications of energy semantic networking [73]. In the energy semantic concept, the system's "big data" enables it to function efficiently by operating with a high level of context awareness. The contextual awareness of the system will guarantee that the shared network between the nodes is capable of interpreting information and user commands as well as communicating them. In addition to data and command processing, the systems elevate user concern when determining the source of power and water by constantly indicting the optimum alternative based on the originator consumption and the varying nodes and main grid availability and pricing. Operating at the community level, the mesogrid enabler will be a semantically capable software, which receives data from the various sensors, devices, user preferences, and other data sources to allow user control over the system's hardware without clashing with the systems operations as well as water and energy consumption patterns. This evolution of the proposed smart water nanogrid provides the following advantages to the resultant smart water/power grid: operational excellence, environmental compliance, grid reliability, safety in operations, energy and water access, security of water and energy supply, consumer participation, grid resilience in normal operation, and disaster situations.

Concluding remarks

In this chapter, a smart water/power technological solution for residential areas has been analyzed based on case study specifications and operating conditions. The solution includes on-site power generation using PV modules, in-pipe hydropower generation from water supply and distribution networks, treatment of gray water via reverse osmosis technology, and reuse of treated gray water at the household level.

Management and control of the water/power technological solution at the household level was done through a centralized controller. Coordination of the water/power components was achieved through networking and communication capabilities facilitated by the controller and GSM technology. This coordination provides the user with real-time data and information about water and power consumptions, flows, and water quality. The user can then make decisions and control actions based on the data and information provided. This allows the user to be in total control of water and power consumptions within their residential area. Although the analysis is based on one case study villa, the same concepts can be applied to other villas and large-scale residential units such as compounds and residential towers without loss of much generality. Experiments with the developed prototype showed that the proposed system is able to (1) generate, store, and provide information that can be used to control water/power consumptions at the household level, (2) allow two-way flows of data and information on the current state of power and water, and (3) treat and recycle treated gray water for use at the villa. The proposed system is expected to reduce freshwater consumption by 20% and power consumption by 25% in residential villas in Qatar. The research study has shown that the in-pipe hydro system can generate small amounts of electricity and contributes to 5% of on-site power generation based on the configurations discussed in this chapter. This contribution is expected to rise in large-scale applications. Payback analysis shows that the combined smart water power nanogrid is moderately attractive and yet environmentally friendly by nature. Prototype tests demonstrated that the proposed system could function properly when implemented in homes. Improvements in gray water collection and treatment processes could result in more benefits. A future improvement of the prototype is to devise the capability to identify the number of leaks as well as determine the exact location of the leaks. Results of such findings can shed light on the further contribution of nanogrids in reducing (a) water losses and (b) water and energy consumptions, thus making homes more energy efficient.

Author details

Dana Alghool, Noora Al-Khalfan, Stabrag Attiya and Farayi Musharavati* Department of Mechanical and Industrial Engineering, College of Engineering, Qatar University, Doha, Qatar

*Address all correspondence to: farayi@qu.edu.qa

References

[1] Marlow DR, Moglia M, Cook S, Beale DJ. Towards sustainable urban water management: A critical reassessment. Water Research. 2013;**47**(20):7150-7161

[2] Asif M, Muneer T. Energy supply, its demand and security issues for developed and emerging economies. Renewable and Sustainable Energy Reviews. 2007;**11**(7):1388-1413

[3] Siddiqi A, Anadon LD. The water–energy nexus in Middle East and North Africa. Energy Policy. 2011;**39**(8):4529-4540

[4] Rao P, Kostecki R, Dale L, Gadgil A. Water-energy nexus: The role of technology and engineering. Annual Review of Environment and Resources. 2017;**42**(1)

[5] WWAP (United Nations World Water Assessment Programme). The United Nations World Water Development Report 2015: Water for a Sustainable World. Paris: UNESCO; 2015

[6] United Nations World Water Day 2014. UN Stresses Water and Energy Issues retrieved October 2017 from https://unu.edu/media-relations/ releases/wwd2014-un-stresses-water- energy-issues.html

[7] DeNicola E, Aburizaiza OS, Siddique A, Khwaja H, Carpenter DO. Climate change and water scarcity: The case of Saudi Arabia. Annals of Global Health. 2015;**81**(3):342-353

[8] The European Electricity Grid Initiative (EEGI) Roadmap 2010-18 and Implementation Plan 2010-12. Presented at SET-PLAN conference; 2010

[9] Li F et al. Smart transmission grid: Vision and framework. IEEE Transactions on Smart Grid. 2010;**1**(2):168-177

[10] Grijalva S, Tariq MU. Prosumer- based smart grid architecture enables a flat, sustainable electricity industry. In: 2011 IEEE PES Innovative Smart Grid Technologies (ISGT). IEEE. 2011, January. pp. 1-6

[11] Mathiesen BV, Lund H, Connolly D, Wenzel H, Østergaard PA, Möller B, et al. Smart energy systems for coherent 100% renewable energy and transport solutions. Applied Energy. 2015;**145**:139-154

[12] Bahrami S, Sheikhi A. From demand response in smart grid toward integrated demand response in smart energy hub. IEEE Transactions on Smart Grid. 2016;**7**(2):650-658

[13] Dominković DF, Bačeković I, Sveinbjörnsson D, Pedersen AS, Krajačić G. On the way towards smart energy supply in cities: The impact of interconnecting geographically distributed district heating grids on the energy system. Energy. 2017

[14] Fadaeenejad M, Saberian AM, Fadaee M, Radzi MAM, Hizam H, AbKadir MZA. The present and future of smart power grid in developing countries. Renewable and Sustainable Energy Reviews. 2014;**29**:828-834

[15] Galo JJ, Macedo MN, Almeida LA, Lima AC. Criteria for smart grid deployment in Brazil by applying the Delphi method. Energy. 2014;**70**:605-611

[16] Cardenas JA, Gemoets L, Rosas JHA, Sarfi R. A literature survey on smart grid distribution: An analytical approach. Journal of Cleaner Production. 2014;**65**:202-216

[17] Siano P. Demand response and smart grids—A survey. Renewable and Sustainable Energy Reviews. 2014;**30**:461-478

[18] Aghaei J, Alizadeh MI. Demand response in smart electricity grids equipped with renewable energy sources: A review. Renewable and Sustainable Energy Reviews. 2013;**18**:64-72

[19] Verbong GP, Beemsterboer S, Sengers F. Smart grids or smart users? Involving users in developing a low carbon electricity economy. Energy Policy. 2013;**52**:117-125

[20] Boroojeni KG, Amini MH, Iyengar SS. Reliability in smart grids. In: Smart Grids: Security and Privacy Issues. Springer International Publishing; 2017. pp. 19-29

[21] Moslehi K, Kumar R. A reliability perspective of the smart grid. IEEE Transactions on Smart Grid. 2010;**1**(1):57-64

[22] Brown RE. Impact of smart grid on distribution system design. In: Power and Energy Society General Meeting- Conversion and Delivery of Electrical Energy in the 21st Century, 2008 IEEE. IEEE. 2008. pp. 1-4

[23] Shao S, Pipattanasomporn M, Rahman S. Grid integration of electric vehicles and demand response with customer choice. IEEE Transactions on Smart Grid. 2012;**3**(1):543-550

[24] Hajebi S, Song H, Barrett S, Clarke A, Clarke S. Towards a reference model for water smart grid. International Journal of Advances in Engineering Science and Technology. 2013;**2**(3):310-317

[25] Lee SW, Sarp S, Jeon DJ, Kim JH. Smart water grid: The future water management platform. Desalination and Water Treatment. 2015;**55**(2):339-346

[26] Spinsante S, Squartini S, Gabrielli L, Pizzichini M, Gambi E, Piazza F. Wireless m-bus sensor networks for smart water grids: Analysis and results. International Journal of Distributed Sensor Networks. 2014;**10**(6):579271

[27] Dawoud MA. The role of desalination in augmentation of water supply in GCC countries. Desalination. 2005;**186**(1-3):187-198

[28] Darwish MA, Al-Najem NM, Lior N. Towards sustainable seawater desalting in the Gulf area. Desalination. 2009;**235**(1):58-87

[29] Khan SUD, Khan SUD, Haider S, El-Leathy A, Rana UA, Danish SN, et al. Development and techno-economic analysis of small modular nuclear reactor and desalination system across Middle East and North Africa region. Desalination. 2017;**406**:51-59

[30] Duan C, Chen B. Energy–water nexus of international energy trade of China. Applied Energy. 2017;**194**:725-734

[31] Wang S, Cao T, Chen B. Urban energy–water nexus based on modified input–output analysis. Applied Energy. 2017;**196**:208-217

[32] Fang D, Chen B. Linkage analysis for the water–energy nexus of city. Applied Energy. 2017;**189**:770-779

[33] Ramachandra T, Ramachandra T, Karunasena G, Karunasena G. Emerging issues in the built environment sustainability agenda. Built Environment Project and Asset Management. 2017;**7**(4):350-352

[34] Rizzo A. Rapid urban development and national master planning in Arab Gulf countries. Qatar as a case study. Cities. 2014;**39**:50-57

[35] Qatar General Electricity & Water Corporation. Statistics Report 2012. Retrieved October 2017, from: https:// www.km.com.qa/MediaCenter/ Publications/ Statisticsreport-2012.pdf

[36] U.S. Environmental Protection Agency. 2009 U.S. Greenhouse Gas Inventory Report. Retrieved October 2016 from: https://www.epa.gov/ ghgemissions/ us-greenhouse-gas-inventory-report-archive

[37] Karlsson P-O, Decker C, Moussall J. Energy efficiency in the UAE: Aiming for sustainability. Retrieved October 2017, from: https://www. strategyand.pwc. com/reports/ energy-efficiency-in-uae

[38] Hodge BK. Alternative Energy Systems and Applications. John Wiley & Sons; 2017

[39] Al-Karaghouli A, Renne D, Kazmerski LL. Solar and wind opportunities for water desalination in the Arab regions. Renewable and Sustainable Energy Reviews. 2009;**13**(9):2397-2407. DOI: 10.1016/j. rser.2008.05.007

[40] Zhang Q , Mclellan BC, Utama NA, Tezuka T, Ishihara K. A Methodology for Designing Future Zero-Carbon Electricity Systems with Smart Grid and Its Application to Kansai Area, Japan. Design for Innovative Value Towards a Sustainable Society. 2012. pp. 50-54. DOI: 10.1007/978-94-007-3010-6_11

[41] Cobacho R, Arregui F, Gascó L, Cabrera E. Low-flow devices in Spain: How efficient are they in fact? An accurate way of calculation. Water Science and Technology: Water Supply. 2004;**4**(3):91-102

[42] Vieira P, Jorge C, Covas D. Assessment of household water use efficiency using performance indices. Resources, Conservation and Recycling. 2017;**116**:94-106. DOI: 10.1016/j. resconrec.2016.09.007

[43] Siddiqi A, de Weck OL. Quantifying end-use energy intensity of the urban water cycle. Journal of Nanogrid Systems. 2013;**19**(4):474-485

[44] Loh M, Coghlan P. Domestic Water Use Study: Perth. Perth, Western Australia: Water Corporation; 2003

[45] Willis RM, Stewart RA, Panuwatwanich K, Capati B, Giurco DP. Gold coast domestic water end use study. Water. 2009;**36**(6):79-85

[46] Beal C, Stewart RA, Huang TT, Rey E. SEQ residential end use study. Journal of Australian Water Association. 2011;**38**(1):80-84

[47] Matos C, Teixeira CA, Duarte AALS, Bentes I. Domestic water uses: Characterization of daily cycles in the north region of Portugal. Science of the Total Environment. 2013;**458**:444-450

[48] Cole G, Stewart RA. Smart meter enabled disaggregation of urban peak water demand: Precursor to effective urban water planning. Urban Water Journal. 2013;**10**(3):174-194. DOI: 10.1080/1573062x.2012.716446

[49] Omaghomi T, Buchberger S. Estimating water demands in buildings. Procedia Engineering. 2014;**89**:1013-1022

[50] Willis RM, Stewart RA, Giurco DP, Talebpour MR, Mousavinejad A. End use water consumption in households: Impact of socio-demographic factors and efficient devices. Journal of Cleaner Production. 2013;**60**:107-115

[51] Matos C, Pereira S, Amorim EV, Bentes I, Briga-Sá A. Wastewater and greywater reuse on irrigation in centralized and decentralized systems—An integrated approach on water quality, energy consumption and CO_2 emissions. Science of the Total Environment. 2014;**493**:463-471

[52] Hunt D, Rogers C. A benchmarking system for domestic water use. Sustainability. 2014;**6**(5):2993-3018. DOI: 10.3390/su6052993

[53] Willis RM, Stewart RA, Panuwatwanich K, Williams PR, Hollingsworth AL. Quantifying the influence of environmental and water conservation attitudes on household end use water consumption. Journal of Environmental Management. 2011;**92**(8):1996-2009

[54] Lee M, Tansel B, Balbin M. Influence of residential water use efficiency measures on household water demand: A four year longitudinal study. Resources, Conservation and Recycling. 2011;**56**(1):1-6. DOI: 10.1016/j. resconrec.2011.08.006

[55] Carragher BJ, Stewart RA, Beal CD. Quantifying the influence of residential water appliance efficiency on average day diurnal demand patterns at an end use level: A precursor to optimised water service nanogrid planning. Resources, Conservation and Recycling. 2012;**62**:81-90. DOI: 10.1016/j. resconrec.2012.02.008

[56] Carr G, Potter RB. Towards effective water reuse: Drivers, challenges and strategies shaping the organisational management of reclaimed water in Jordan. The Geographical Journal. 2013;**179**(1):61-73

[57] Gourbesville P. Why smart water journal? Smart Water. 2016;**1**(1):1-2

[58] Mutchek M, Williams E. Moving towards sustainable and resilient smart water grids. Challenges. 2014;**5**(1):123-137

[59] Harris R. WaterSmart Toolbox: Giving customers online access to real-time water consumption is the right tool for water efficiency success. In: Presented at the Smart Water Innovations Conference and Exposition, 6-8 October; Las Vegas, NV. Available from: http://watersmartinnovations.com/2010/ PDFs/10-T-1001.pdf. 2010 [Accessed: 10 October 2016]

[60] Weng KT, Lim A. Pursuit of a smart water grid in Singapore's water supply network. In: Presented at the American Water Works Association Sustainable Water Management Conference, 18-21 March; Portland, OR. 2012

[61] Alghool D, Al-Khalfan N, Attiya S. Design of a smart water/power system for households. Project Report. Qatar University; 2016

[62] Sai Y, Cohen S, Vogel RM. The impacts of water conservation strategies on water use: Four case studies. Journal of the American Water Resources Association (JAWRA). 2011;**47**(4):687-701. DOI: 10.1111/j.1752-1688.2011.00534.x

[63] Gabbar HA. Engineering design of green hybrid energy production and supply chains. Environmental Modelling & Software. 2009;**24**(3):423-435

[64] Gabbar HA, Bondarenko D, Hussain S, Musharavati F, Pokharel S. Building thermal energy modeling with loss minimization. Simulation Modelling Practice and Theory. 2014;**49**:110-121

[65] Casini M. Harvesting energy from in-pipe hydro systems at urban and building scale. International Journal of Smart Grid and Clean Energy. 2015;**4**:316-327

[66] Porkumaran K, Tharu RP, Sukanya S, Elezabeth VV, Gowtham N. Micro in-pipe hydro power plant for rural electrification using LabVIEW. In: 2017 International Conference on Innovations in Green Energy and Healthcare Technologies (IGEHT); IEEE. 2017. pp. 1-5

[67] Rakesh C, Nallode C, Adhvaith M, Anwin TJ, Krishna AA, Rakesh C, et al. Theoretical study and performance test of lucid spherical turbine. International Journal. 2016;**3**:418-423

[68] Haidar AM, Senan MF, Noman A, Radman T. Utilization of pico hydro generation in domestic and commercial loads. Renewable and Sustainable Energy Reviews. 2012;**16**(1):518-524

[69] Williams AA, Simpson R. Pico hydro-reducing technical risks for rural electrification. Renewable Energy (Elsevier). 2009;**34**:1986-1991

[70] Smits M, Bush SR. A light left in the dark: The practice and politics of pico-hydropower in the Lao PDR. Energy Policy (Elsevier). 2010;**38**:116-127

[71] Abbasi T, Abbasi SA. Small hydro and the environmental implications of its extensive utilization. Renewable and Sustainable Energy Reviews (Elsevier). 2011;**15**:2134-2143

[72] Fodorean D, Member SL, Miraoui A. Generator solutions for stand alone pico-electric power plants. Proceedings of IEEE Conference on Electric Machines and Drives. 2009:434-438

[73] Gabbar HA, Eldessouky AS. Energy semantic network for building energy management. Intelligent Industrial Systems. 2015;**1**(3):213-231

7

Resonant Power Converters

Mohammed Salem and Khalid Yahya

Abstract

Recently, DC/DC resonant converters have received much research interest as a result of the advancements in their applications. This increase in their industrial application has given rise to more efforts in enhancing the soft-switching, smooth waveforms, high-power density, and high efficiency features of the resonant converters. Their suitability to high frequency usage and capacity to minimize switching losses have endeared them to industrial applications compared to the hard-switching conventional converters. However, studies have continued to suggest improvements in certain areas of these converters, including high-power density, wide load variations, reliability, high efficiency, minimal number of components, and low cost. In this chapter, the resonant power converters (RPCs), their principles, and their classifications based on the DC-DC family of converters are presented. The recent advancements in the constructions, operational principles, advantages, and disadvantages were also reviewed. From the review of different topologies of the resonant DC-DC converters, it has been suggested that more studies are necessary to produce power circuits, which can address the drawbacks of the existing one.

Keywords: soft-switching, LLC resonant converter, control strategies, resonant power converters (RPCs),

Introduction

Several studies have been conducted on the switching mode DC-DC converters to ensure that they satisfy the most demanding criteria for power electronics application. The possibility of minimizing the switching and conduction losses in the switch-mode through increasing the switching frequency makes them more attractive. However, several switching topologies can attain a high-power transfer [1, 2] but the problem is the power switches (transistors or MOS-FET), diodes, and energy storage passive elements (capacitors and inductors) contained in the structure of the power converters, which affects their efficiency. Efficient circuits have been developed for power converters along with the developments in materials, control systems, and devices technology. Such topologies can minimize the overall converter size and the switching losses, thereby, providing a higher efficiency [3–5]. The DC-DC converters can be categorized into three major groups (linear, hard- switching, and soft-switching mode resonant converters) based on their modes of operation as depicted in **Figure 1** [6–8].

The advantages of the linear regulator technologies include low noise, fast response, simplicity, and excellent regulation. However, they have some disadvantages such as power dissipation in any working condition, which can result in low efficiency. The switching-mode topologies can be categorized based on the isolation feature to galvanic isolation converter, which also called chopper, and non-isolated converter. Galvanic isolation is required in most applications during the process of converting the power from the utility grid (voltages ranging from 100 to 600 V) for safety reason [9–11]. Since a higher power transformer is required in these converters, the single-switch converter may not be a proper solution for higher power applications, and as such, the other DC-DC isolated converters with more than one switch, such as the push-pull and full or half-bridge converters, are more appropriate for such high-power applications. Hard-switching transitions in devices operating with switched-mode converters will result in high power dissipation and a gradual reduction of the efficiency of the converter; it can damage the switching devices as well. Snubbers are used to reduce the stress of the switches and solve this problem of power dissipation. Furthermore, the hard-switching converters have other disadvantages such as limited frequency, high EMI, high switching losses, large size, and heavy weight. Two more problems are encountered during the control of the transferred power; the first one is the generated noise during the switching process while the second one is the energy lost in the switches.

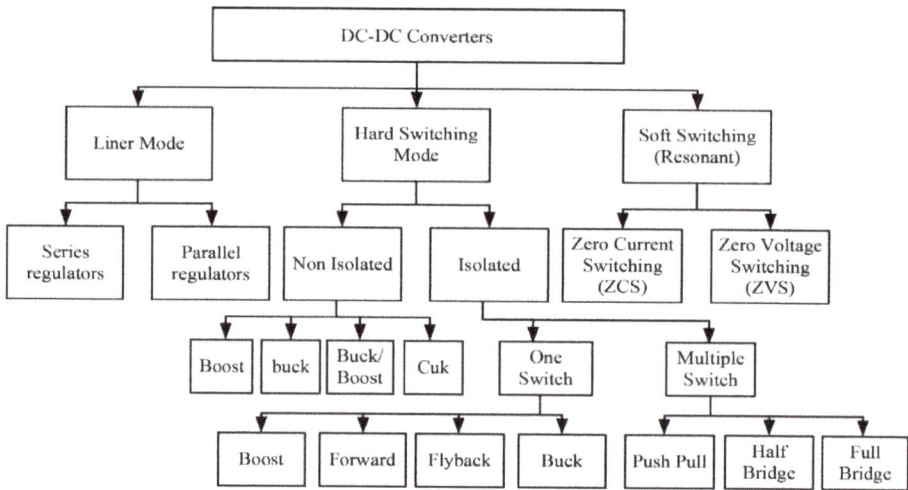

Figure 1. *The classification of the DC-DC converters.*

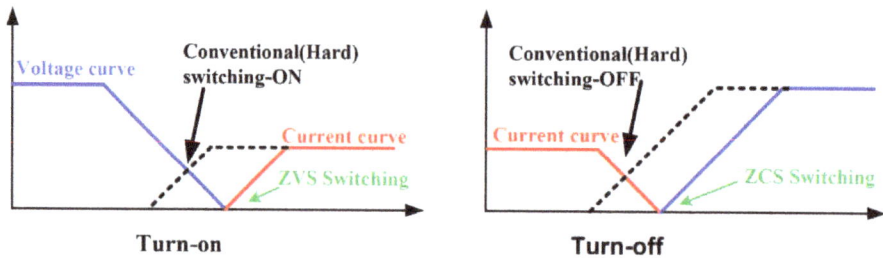

Figure 2. *Current waveform and voltage waveform of hard-switching and soft-switching at both turn-on and turn-off transitions.*

Regardless of these drawbacks, this study implemented the hard-switching transitions [12]. The voltage and the active switches' current were modified to overcome or minimize their effects [13]. These modified methods work by either forcing the voltage across the switch or current through the switch to zero, and such a transition technique can only be achieved with a soft-switching technique (**Figure 2**). The current waveform is forced to reduce to 0 in the zero current switching (ZCS) circuit while the voltage waveform is treated as such by the zero voltage switching (ZVS) circuit. Power circuits with these types of transition techniques are called soft-switching converters (SSCs) [14]. Meanwhile, there are several advantages of the SSCs over the linear mode regulators. The advantages include (i) the possibility of using a small ferrite transformer core due to the high- switching frequency, making operation in a wider DC input voltage range possible compared to the linear regulators; and (ii) it offers a higher efficiency. However, the complexity of the control circuit is associated with several drawbacks compared to the linear control circuit; and the power switching technique may increase the supply noise.

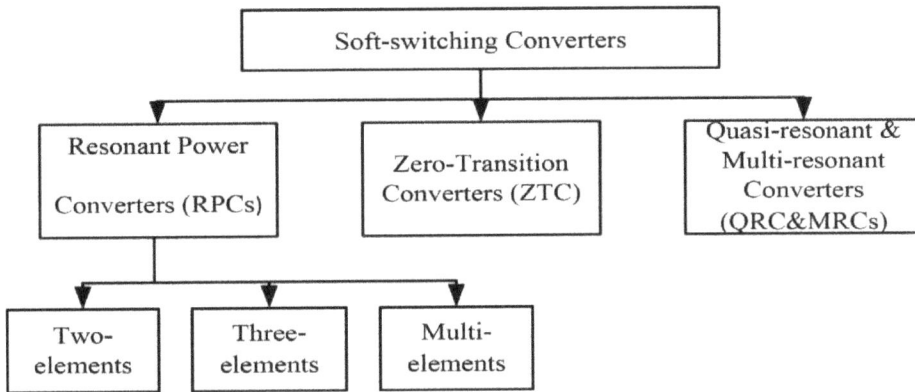

Figure 3. *Categories of the SSCs.*

The SSCs have been significantly improved in the areas of switching losses, the EMIs, and device stresses, allowing the converters to perfectly work even at higher frequencies and a consequent reduction in the magnetic components. Generally, the SSC families are categorized into resonant transition converters (RTCs), resonant power converters (RPCs), quasi-resonant converters (QRCs), and multi-resonant converters (MRCs) based on their modes of operation, as shown in **Figure 3**.

Resonant power converters (RPCs)

Figure 4 shows the structure of the RPCs and each stage represents a specific job to be carried out. The controlled switching network (CSN) is powered by the DC source; it rapidly switches on and off depending on the working frequency to generate the output voltage or current, which feeds the next stage. The sinusoidal voltage and current signals are generated at a stage of the high-frequency resonant tank network (RTN), where there are two or more reactive components. This is to ensure a reduced electromagnetic interference and harmonic distortion [15]. Being deployed as an energy cushioning stage between the load and the CSN, a frequency- selective network can identify this stage. The impedances of both capacitance and inductance at resonance condition are equivalent and will produce the resonant frequency. A rectifier network and a pass filter are then used for rectifying and filtering the output signal to generate the anticipated DC output voltage [16].

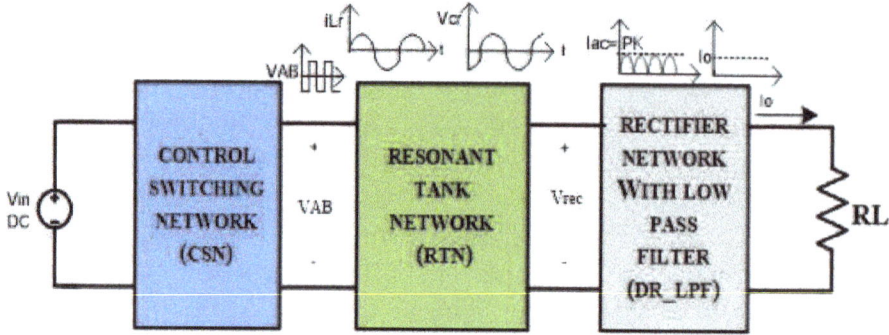

Figure 4. *The structure of the RPC.*

Control switching network (CSN)

The full and half bridge are the commonest switching networks, whose usage is power-dependent. For high-power applications, the full bridge inverter is often used as opposed to the single ended or half bridge inverters, which can only supply the active switch with half of the input voltage. This indicates that full bridge inverters have a low rate of voltage switch, making them ideal for application in high input voltage conditions [17]. The RPCs and either a half or full-bridge inverter are often deployed together along with each of center-tapped or full-bridge rectifiers [18].

The CSN depicted in **Figure 5** generates a square waveform voltage Vs(t) of the switching frequency fs ($\omega_s = 2\pi$ fs) as represented in Eq. 1 by the Fourier series. Considering the response of the resonant tank, which is dominant to the basic constituent fs of the voltage waveform Vs(t), then, the infinitesimal response clearly demonstrated the harmonic frequencies nfs, n = 3, 5, 7, ... As a result, the power that correlates to the basic voltage waveform constituent Vs(t) is moved to the resonant tank as represented in Eq. (2). The basic constituent is a sinusoidal waveform with a peak amplitude of (4/π) times the DC source voltage. This basic constituent and the original waveform are in the same phase. However, when the S1 is on, there is a positive output sinusoidal switched current (t) but negative when S2 is off. This is due to the alternate working principle of the two switches, and its peak amplitude Is1 with phase equal to φs. The input current (DC) to the CSN can be computed by dividing the sinusoidal switched current with half the switching period [6, 16].

$$V_s(t) = \frac{4V_g}{\pi} \sum_{n=1, 3, 5, \dots} \frac{1}{n} \sin(n\omega_s t) \tag{1}$$

$$V_{s1}(t) = \frac{4V_g}{\pi} \sin(n\omega_s t) \tag{2}$$

$$i_s(t) = I_{s1} \sin(\omega_s t - \varphi_s) \tag{3}$$

$$I_{in} = \frac{2}{T_s} \int_0^{\frac{T_s}{2}} i_s(t) d_t = \frac{2}{\pi} I_{s1} \cos(\varphi_s) \tag{4}$$

Resonant tank network (RTN)

Resonant tank networks (RTNs) comprise of LC circuit (reactive elements) that stock oscillating energy with the frequency of circuit resonant. The LC circuit's resonance h attains the electromagnetic frequency useful in several applications, including the telecommunication technology. The tank can possibly be charged to a certain resonant frequency through the adjustment of reactive element data. In addition, the essential phase of the resonant power converter is the resonant tank network. There are different kinds of RTN; it is mainly categorized into three parameters. Firstly, the resonant power converter can be sectioned through the connection technique used in tank element. The main common three resonant circuits include a series-parallel resonant converter (SPRC), a series resonant converter (SRC), and parallel resonant converter (PRC) [19]. The second factor lies in a quantity of the reactive elements (amount of transfer function order). However, the third one depends on the elements and multi-elements resonant tank [16].

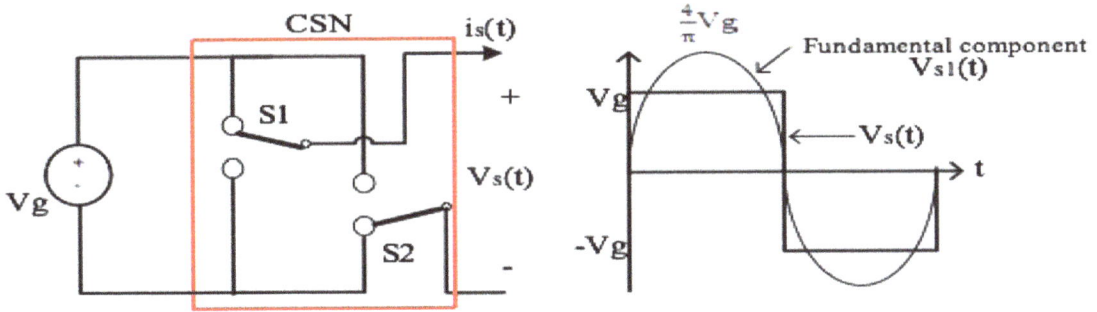

Figure 5. *The equivalent circuit of CSN.*

Topographies of the three elements RTN (third order resonant tank) are controlled in overpowering the inadequacies in the two elements RTN. Most especially, the third element is put in the two elements RTN with a certain rumination to generate the three elements RTN. Thus, these can be taken as an integration of the advantages of mostly used two elements resonant converters PRC and SRC and enhance their inadequacies. The third order RTN contains 36 various tanks, the most common used tanks are LCC, LCL, and CLL [14, 20]. Multi-element resonant converters are RTN that contains four and beyond the number of elements. It should be noted that an increase in the number of reactive elements cause the network to be more complicated based on its size, analysis, and cost. For instance, **Figure 6** illustrates the fourth order RTN, which is referred to as the LCLC tank system [21]; this topology includes the characteristics of two main famous three-element systems such as LLC and LCC, and then reflects on their setbacks.

Diode rectifier network with low pass filter (DR-LPF)

The RTN generates voltage waveforms and sinusoidal current based on the output voltage and resonant frequency, which are taken to be the input to the last phase DR-LPF is the pulse waveform. This implies that the purpose of utilizing DR- LPF is to filter and correct the AC waveform to achieve the entailed DC output waveform. In the previous studies on resonant power converter, the DR-LPS had been illustrated as a full-bridge or center-tapped rectifiers. While the center-tapped rectifier is inappropriate due to an elevated voltage stress from the diodes, the low pass filter had been investigated for the entire occurrences of inductance or capacitive [16, 22].

Diode rectifier with capacitive low pass filter (DR-CLPF)

In the DR-CLPF, the input voltage VR(t) is appraised as the square wave of a resonant frequency as illustrated in **Figure 7**. The input voltage VR(t) can be estimated based on the resonant tank filtering with its basic component VR1(t) as shown in Eqs. (6) and (7), respectively. In addition, the basic component and current are in the same phase, any drop in current to zero alters the basic compartment because of the variations in the conducting diodes.

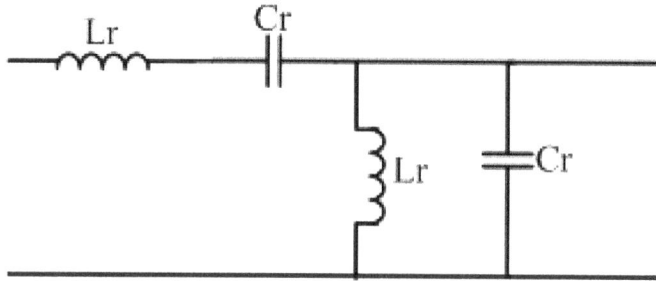

Figure 6. *An illustration of four-element RTN (LCLC).*

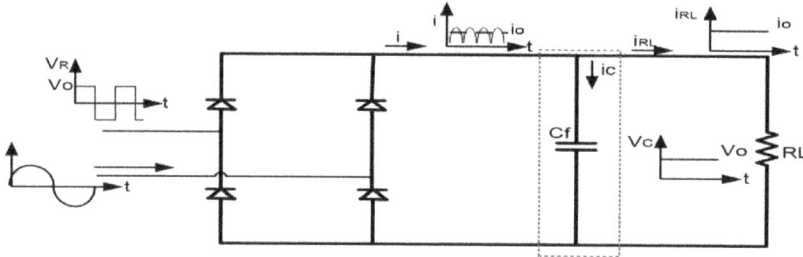

Figure 7. *DR-CLPF containing a capacitive pass filter and its variables waveforms.*

$$i_R(t) = I_p \sin(\omega_s t - \varphi_s) \tag{5}$$

$$V_R(t) = \frac{4V_o}{\pi} \sum_{n=1,\,3,\,5,\,\dots} \frac{1}{n} \sin(n\omega_s t - \varphi_s) \tag{6}$$

$$V_{R1}(t) = \frac{4V_o}{\pi} \sin(\omega_s t - \varphi_s) \tag{7}$$

$$I_o = \frac{2}{T_s} \int_0^{\frac{T_s}{2}} i_R(t) d_t = \frac{2}{\pi} I_R \tag{8}$$

Diode rectifier coupled with inductive low pass filter (DR-LLPF)

The diode rectifier is coupled with the input voltage VR(t) (sinusoidal wave- form) and an inductive filter jacket. The current inputted is appraised as the square waveform iR(t) as illustrated in **Figure 8**.

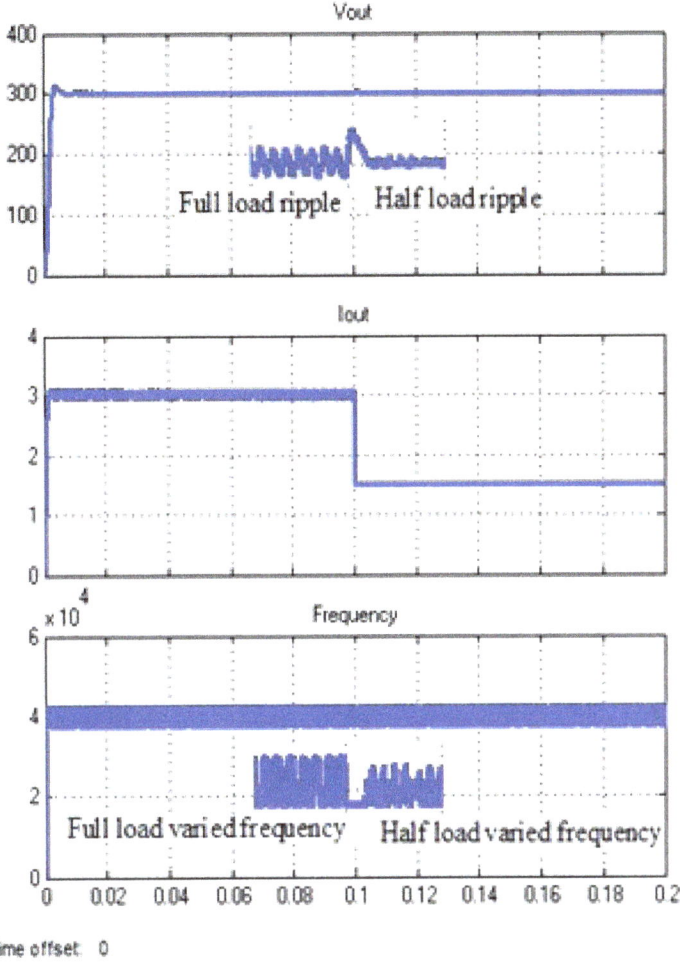

Figure 8. *DR-LLPF coupled with an inductive pass filter and its variables waveforms.*

$$V_R(t) = V_p \sin (\omega_s t - \varphi_s) \qquad (9)$$

$$i_R(t) = \frac{4I_o}{\pi} \sum_{n=1, 3, 5, \dots} \frac{1}{n} \sin (n\omega_s t - \varphi_s) \qquad (10)$$

$$i_{R1}(t) = \frac{4I_o}{\pi} \sin (\omega_s t - \varphi_s) \qquad (11)$$

$$V_o = \frac{2}{T_s} \int_0^{\frac{T_s}{2}} V_R(t) d_t = \frac{2}{\pi} V_R \qquad (12)$$

Properties of resonant power converters

Parallel resonant converter (PRC)

The PRC is categorized into two elements tank converter. The resonant capacitor Cr needs to be parallel to the diode rectifier network DR and load. For the effective load resistance, Rac is much more enormous relative to resonant capacitor reactance Cr; this implies that the resonant current is unproportionable to the load. Moreover, in addition, the voltage over the resonant capacitor and parallel resistance Rac can be improved by declining the load [14]. **Figure 9** shows the relationship between the load quality factor and PRC voltage gain depending on switching frequency based on Eq. (13). It can be observed that the high voltage gain is attained from the switching frequencies and light load conditions, which are nearly equivalent to resonant frequency fs = fr. Therefore, PRC can either step the output voltage down or up depending on the variation in the control switching system frequency. The voltage output can be adjusted with load states, whereas, the resonant current is restricted to resonant inductor data, this causes the PRC to be appropriate for open and short circuit applications [23].

$$M_V = \frac{1}{\sqrt{\left[1 - \left(\frac{f_s}{f_r}\right)^2\right]^2 + \left[\frac{f_s}{f_r}\left(\frac{1}{Q}\right)\right]^2}} \tag{13}$$

LLC resonant converter

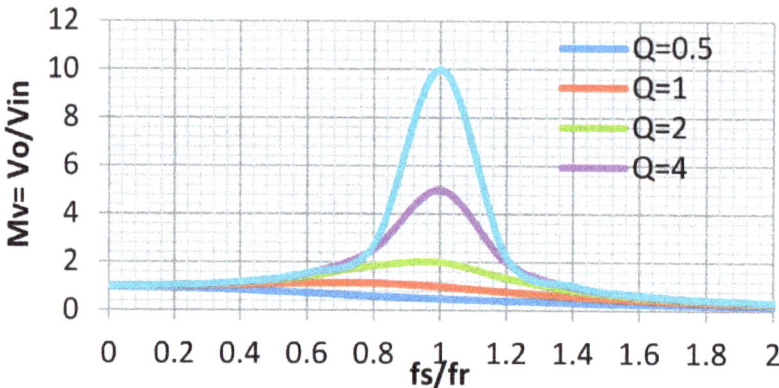

Figure 9. *Voltage gain characteristic of PRC.*

LLC resonant converter in the RTN comprises three reactive elements, whereby it is appraised as a conventional SRC and addition of inductor L_r equidistant to the load that is referred to as parallel inductor L_{rp}. The parallel inductor can be replaced by using the magnetizing inductance when using a transformer [14, 16]. The topology of LLC generates two resonant frequencies: firstly, the series resonant frequency fr1 depending on the series elements $L_r C_r$ and secondly, the parallel resonant frequency fr2 depending on the entire three tank elements (L_{rp}, C_r, and L_r) as illustrated in Eqs. (14) and (15), where $f_{r1} > f_{r2}$.

$$f_{r1} = \frac{1}{2\pi\sqrt{(L_r + L_{rp})C_r}} \tag{14}$$

$$f_{r2} = \frac{1}{2\pi\sqrt{L_r C_r}} \tag{15}$$

Figure 10 illustrates the LLC voltage gain depending on the unity inductance ration (AL = 1), switching frequency, and load quality factor as shown in the Eq. (16).

$$M_V = \cfrac{1}{\sqrt{\left[1 - \left(\frac{f_s}{f_r}\right)^2\right]^2 + \left[\frac{f_s}{f_r}\left(\frac{1}{Q}\right)\right]^2}} \tag{16}$$

LCC resonant converter

The topology of LCC in the RTN comprises three reactive elements, the capacitor Crp that is connected to the load in parallel represents the third element. Thus, the topology contains two resonant frequencies: firstly, the series resonant frequency fr1 depending on the series elements Lr Cr and secondly, the parallel resonant frequency fr2 depending on the entire three tank elements (Lrp, Cr, and Lr) as illustrated in as they are shown in Eqs. (17) and (18), where fr1 < fr2.

$$f_{r1} = \frac{1}{2\pi\sqrt{L_r C_r}} \tag{17}$$

$$f_{r2} = \frac{1}{2\pi\sqrt{L_r\left(\frac{C_r C_{rp}}{C_r + C_{rp}}\right)}} \tag{18}$$

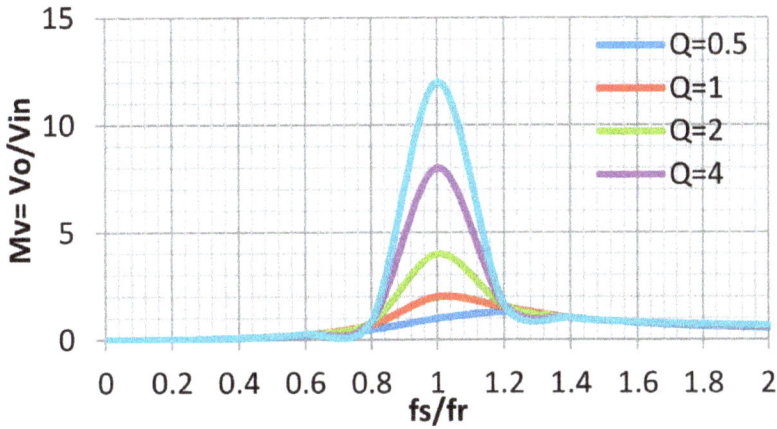

Figure 10. *Voltage gain property of LLC converter.*

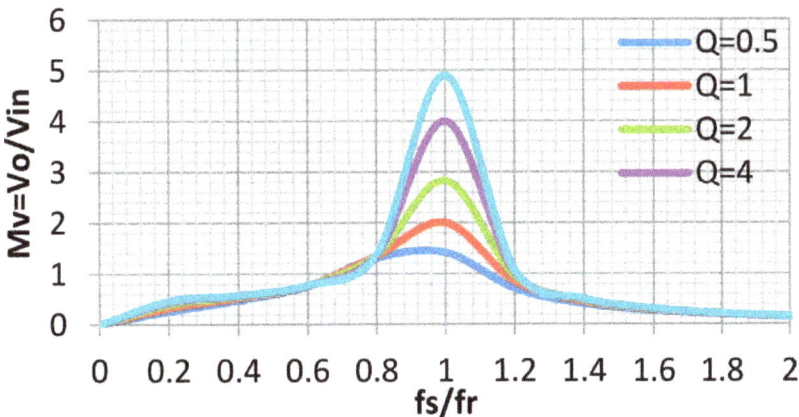

Figure 11. *Voltage gain property of LCC converter.*

In the LCC converters, the proportion of resonant capacitors AC should be chosen prudently to be equal to the targeted peak gain. **Figure 11** explains the LCC voltage gain as depending on the load quality parameter and switching frequency with single inductance ratio (AC = 1) as described in Eq. (19). It is seen that the light load voltage gain advances across the converter properties and parallel resonant frequency fr2 acts as a parallel resonant converter PRC.

$$M_V = \frac{1}{\sqrt{(1+A_C)^2 \left[1 - \left(\frac{f_s}{f_{r2}}\right)^2\right]^2 + \left[\frac{1}{Q}\left(\left(\frac{f_s}{f_{r1}}\right) - \frac{A_C}{A_C+1}\frac{f_{r1}}{f_s}\right)\right]^2}} \qquad (19)$$

LCLC resonant converter

Similar to LCC, the topology of LCLC in the RTN comprises four reactive elements as illustrated in **Figure 6**, where this topology homogenizes the characteristics of LLC and LCC. The topology structure comprises parallel resonant capacitor Crp, parallel resonant inductance Lrp, and a series elements Lr Cr, which implies that the topology contains two proportions whereby the inductance AL and capacitance AC must be appraised in the design. Moreover, the topology comprises three frequencies: two parallel frequencies frp1, frp2, and one series resonant frequency frs. **Figure 12** reflected the relationship between the load quality factor, AC = AL = 1 and LCLC voltage gain depending on the switching frequency as shown in Eq. (20).

$$M_V = \frac{1}{\sqrt{\left[1 + A_C + A_L - A_C\left(\frac{f_s}{f_r}\right)^2 - A_L\left(\frac{f_r}{f_s}\right)^2\right]^2 + \left[\frac{1}{Q}\left(\frac{f_r}{f_s} - \frac{f_s}{f_r}\right)\right]^2}} \qquad (20)$$

Controlled strategies of resonant converters

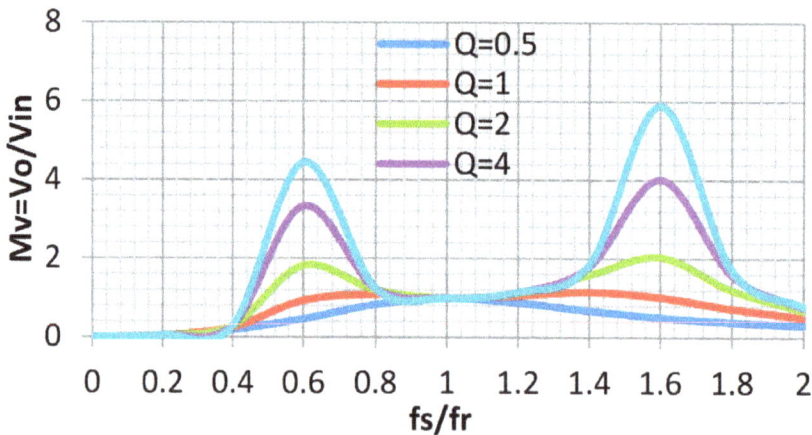

Figure 12. *Voltage gain property of LCLC converter.*

The controlled strategies of the resonant converter is a bit disparate from the pulse width modulated (PWM) converters. A lot of parameters should be considered to attain a soft switching at a certain segment to fabricate the precise controller that can achieve the desired results like load condition, energy storage elements, frequency range, and among others. There were several controlled topologies, which had been applied in the previous studies to manage the series resonant converters. For example, the pulse density modulation, voltage and current control, diode conduction control, and frequency control [24]. The full bridge resonant converter voltage had been regulated through the phase shift control; this phenomenon can be referred to as the switching signal primary control. In addition, to enhance the outcome of a control system, several improved techniques through adaptive controls had been reported [25], which include the passivity-based control and auto disturbance-rejection control (ADRC). A phase shift control had been used to control the current of resonant [26]. Because of this, the control outcome was increased in relative to the traditional PSRC control system. From the previous studies, the controlled techniques can be categorized into their implementation technique either through analog or digital. The digital controls are used because of their flexibility features in compact, programming, and light in comparison to analog controllers, they are also more resistant to inferences and noise. A three element DC-DC resonant converter type LLC has been discussed in this part, in order to compare the performance of frequency (duty-cycle) with variable frequency control, in terms of wide load variation.

Moreover, to ensure the expansion range of ZVS for entire inverter switches (S1–S4), and to improve the converter voltage gain. Then, the magnetizing inductance and resonant tank are used to generate a second resonant frequency $f_{r1} = \frac{1}{\sqrt{(L_r + L_m).C_r}}$.

From **Figure 14**, the LLC resonant converter gain M is evaluated using voltage divider law by considering the load quality factor Q and transformer step-up ratio as shown in Eq. (21).

$$M = \frac{V_o}{V_{in}} = \frac{Z_i}{Z_o} = \frac{F^2(A_L - 1)}{n\sqrt{\left(\frac{f_s^2}{f_{r2}^2} - 1\right)^2 + \left(Q(F^3 - F)^2(A_L - 1)\right)^2}} \tag{21}$$

Fixed frequency control

The purpose of using this technique is that the voltage output is being influenced by changing the duty-cycle to attain a targeted voltage output. However, the error that exists between the constant desired voltage (reference voltage) $V_{err} = V_{ref} - V_{meas}$ and measured voltage is used for the PI controller. Thus, the controlled signal pertained to the PWM generator through the utilization of the switching frequency in generating a gate signal for the entire switches. From Eq. (21), the switching frequency that is equal to 42.5 kHz is the highest gain that can be obtained if the switching frequency and resonant frequency are closer. Therefore, a duty cycle is the only factor that measures the output voltage. Because the voltage ripple can vary by changing the load, then, the duty cycle will react to the variation of load variation relative to the estimated output voltage ripple.

Variable-frequency control

This control varies the switching frequency in adjusting the output voltage so as to reach the targeted load stage and save the output voltage of converter stable in any situation as illustrated in **Figure 5**. Moreover, the error that exists between the stable targeted voltage (reference voltage) $V_{err} = V_{ref} - V_{meas}$ and measured is used in the PI controller. Therefore, the controller increases or decreases the switching frequency based on the targeted output voltage if any variation occurs from the required output or error sign. Thus, a signal will be sent to the entire device switches. Depending on the magnetizing inductance and resonant impedance, the tank response needs to be saved inductively to appraise the attainment of the ZVS in the entire switches. However, the ranges of controlled switching frequency need to be restrained by higher and lower frequencies that affirm the fulfillment of ZVS (45 - 37.5 kHz).

Simulation results

The model simulation of resonant converter is carried out and implemented by utilizing the MATLAB/SIMULINK software using the factors enlisted in **Table 1**; this is to verify the LLC series resonant DC-DC converter design circuit as shown in **Figure 13** and control techniques analysis.

In this frequency control, the controlled signal is used for the PWM generator in generating the gate signals for the entire switches. Then, the switches S_1 and S_4 are enhanced concurrently and replace the S_2 and S_3 switches to generate the input voltage for the resonant tank V_{AB}. Therefore, the resonant elements generate the voltage V_{Cr} and sinusoidal current i_{Lr} as illustrated in **Figure 15**.

Table 1. *The parameters of the LLC converter.*

V_{in}	100 (V)
L_r	200 (µH)
C_r	70 (nF)
C_f	4.7 (µF)
V_o	300 (V)
R_L	100 (Ω)
L_m	300 (µH)
Kp, Ki Duty-cycle control	2.7, 0.5
f_s ranges for VF control	37.5 – 45 (kHz)

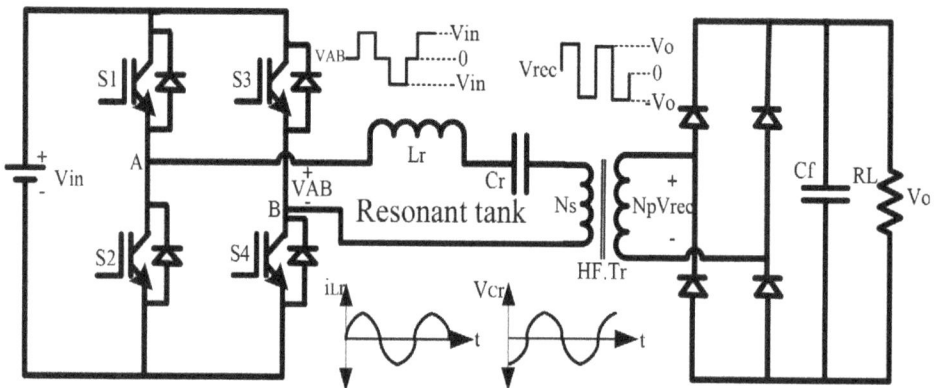

Figure 13. *A simplified illustration of full-bridge LLC resonant converter.*

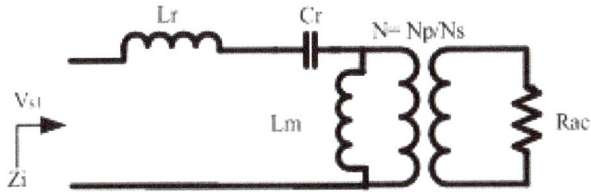

Figure 14. *The AC equivalent circuit between the rectifier and inverter.*

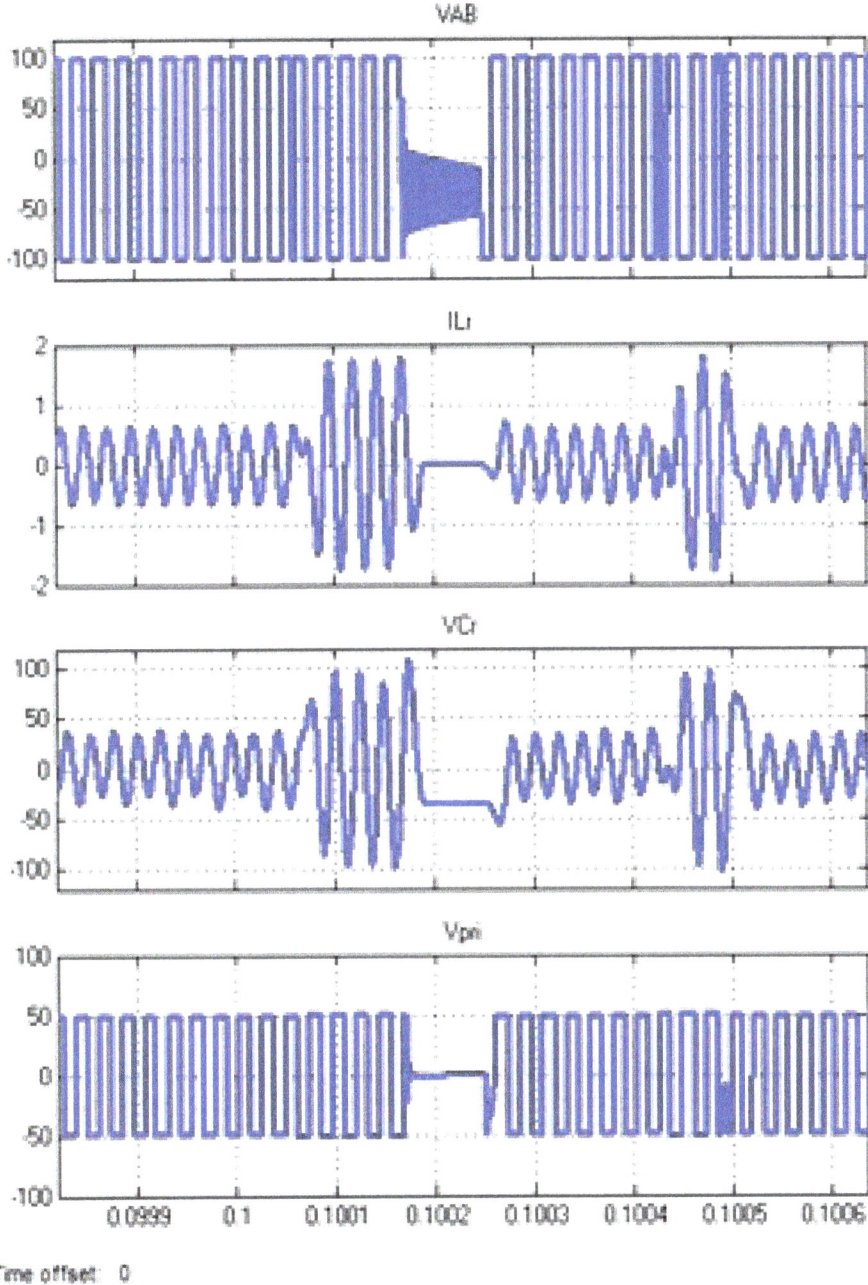

Time offset: 0

Figure 15. *Simulation waveforms of the resonant tank input voltage V_{AB}, resonant inductor current i_L, resonant capacitor voltage V_{cr}, and transformer primary voltage V_{pri}.*

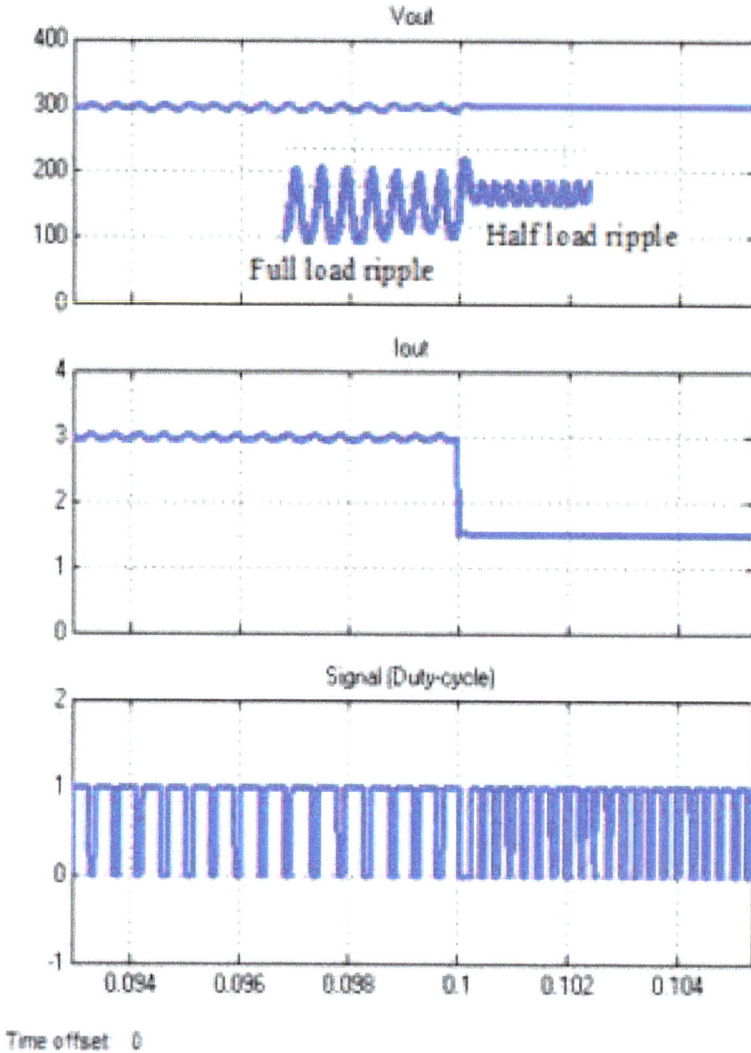

Figure 16. *The dynamic response of the output voltage, output current, and controlled duty-cycle signal with respect to the load changes.*

Figure 16 illustrates an output voltage of the duty-cycle controller dynamic response, the load is stepped up and down within half load of (200 Ω) and a full load of (100 Ω). It was observed that the output voltage ripple is enormous at full load condition in relative to the half load state that reflects a direct duty cycle changes within the entire load conditions. Although, the system generates a favorable outcome by controlling the output voltage equivalent to 300 V, nevertheless, the resonant tank parameters mislaying the resonant concept during the changes in load changes as illustrated in **Figure 15**. At simulation time t = 0:1 s, the load varies the AC tank parameters hold-up by the 1:5e - 3 s, thereafter, the resonant parameters reproduce the targeted AC parameters depending on a value of the load, which can be appraised as a disadvantage of the duty-cycle control technique.

In the variable frequency control technique, the measured output voltage is used to detect frequency. Thereafter, the controlled signal is implemented on PWM to produce gate signals to the entire switches by considering the switching duration depending on the converter nature to generate enormous output voltage gain. Moreover, the variation in load is being tested and applied to affirm the controller dynamic responses. **Figure 18** illustrates the parameter frequency controller dynamic response of the output voltage, and the load is stepped up and down in the similar way of duty-cycle control. It can also be observed that in this control technique, the full load condition results in enormous output voltage ripple as compared to half load state, and this affirms the significant variation in frequency with the entire load condition. Moreover, the parameter frequency control gives a significant response to the tank AC parameters as illustrated in **Figure 17**. As the load varies, the AC parameters temporarily react by increasing or decreasing the voltage and resonant current values depending on the varied frequency and keeping a shape of the sinusoidal waveform.

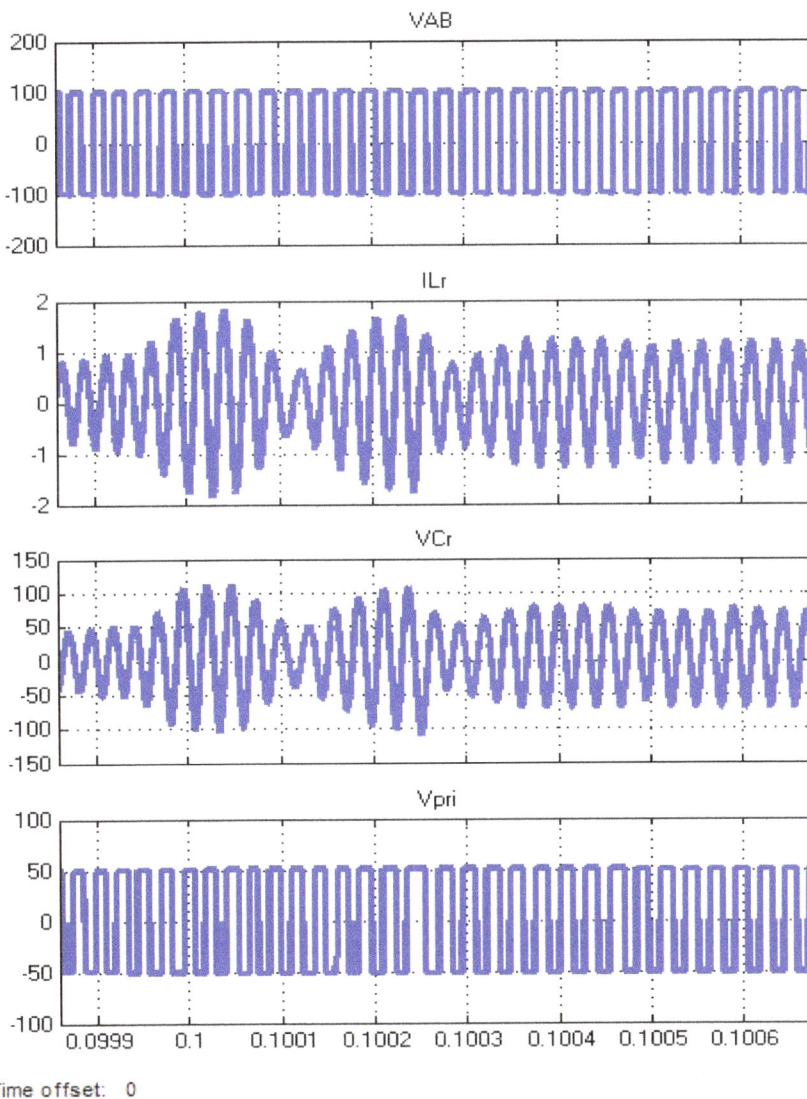

Time offset: 0

Figure 17. *Simulation waveforms of the resonant tank input voltage V_{AB}, resonant inductor current i_L, resonant capacitor voltage V_{cr}, and transformer primary voltage V_{pri}.*

Applications of resonant power converters

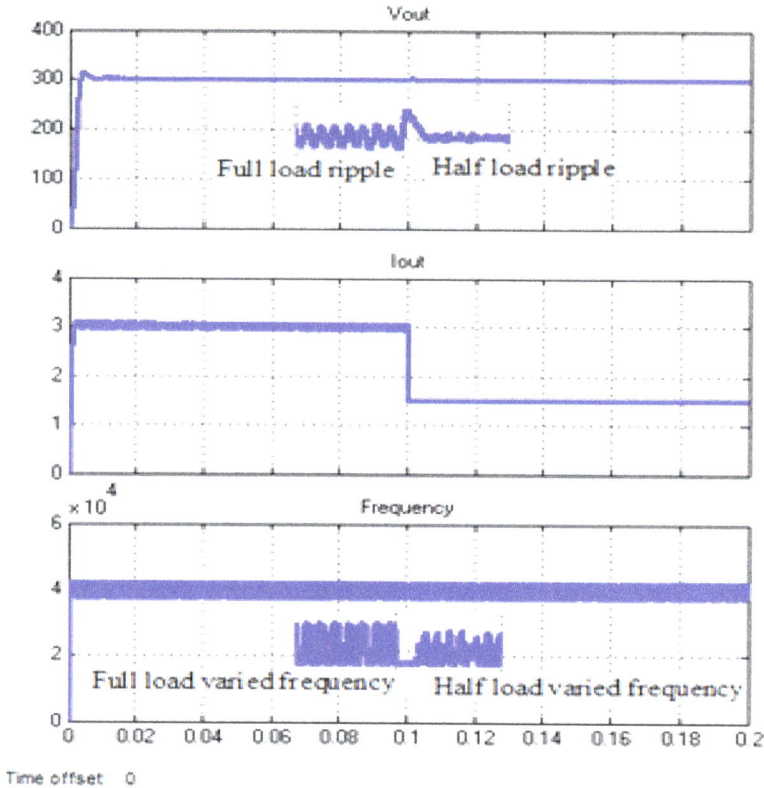

Figure 18. *The dynamic response of output voltage, output current, and controlled frequency signal as the function of load change from full to half load.*

Based on the sufficient demerit of RPCs as earlier stated in the above segments, they have uncommon application in modern industries. The summary of the noticeable implementations is discussed in this segment. The main areas of RPCs application are household applications like induction cookers, portable power supplies, network connection of renewable energy mains, and hybrid and electric vehicles. In a case of the portable power supply, requirements of the converter include a low price tag, light-weight and small size, high efficiency, high reliability, and low electromagnetic interference (EMI). Soft switching is the way of ensuring higher efficiency; it can be implemented by utilizing RPCs. Based on the area of application, the topology can be chosen to ensure maximum efficiency, ideal cost, and size. For example, the supply of power to an electron beam welding compartment uses a full bridge LLC resonant converter [16]. The soft switching technique and topology solves the problem associated with power utilization within the filament supply by staying away from the inverter heating challenge and ensures higher efficiencies. RPCs are used in the electrostatic precipitator. This is a high- power appliance industrially utilized for removing smoke and dust from a flowing gas. The series-parallel RPC coupled with phase control suggested by [27] is negligible in size; it gives a faster temporary response and possesses a higher efficiency as compared to the traditional line frequency power supplies. RPCs are known for charging hybrid vehicles whereby the batteries need to be charged either by wireless or wired. Due to being smaller and demonstrating higher effectiveness, the charging of EVs through wire [28] and plug-in EVs [29] RPCs are used. For example, the high-performance LLC converters are suggested in the two-stage smart battery chargers. The converter evacuates the low and frequencies ripples from the current output and increases the life of a battery without the size increment of the charger. Previous studies had proposed several wired charging topologies. Apart from wired chargers, wireless power transfer (WPT) has been a modern charging process used in the hybrid and electric vehicles. The accessible WPT technologies include electromagnetic, magnetic, and electric power transfer. Among all of these, the magnetic coupling method utilizes RPCs, which portray higher power transmission and higher efficiency within a closer distance. Application of RPC in WPT for hybrid and electric vehicles had been reported [30]. The grid integration of renewable energy mains such as fuel cell, wind, and solar PV needs converters of minimal current ripple and higher efficiency. DC-DC converters are the main prerequisite in processing power from renewable energy mains. Out of several options, RPCs can be the main competitor because of its low EMI, higher efficiency, robustness, and low output ripple current. RPCs are used in FC networks [31, 32], PVs [33], grid connection, and electrolyser interfaces [34].

Moreover, RPCs is applied to home induction cookers. The resonant inverter is the induction cooker key element; it produces an AC current that heats up the inductor- vessel compartment. The resonant inverters utilize in induction cooker are multilevel, half bridge, single-switch, and full bridge inverters [35].

Conclusion

This chapter has vividly explained the resonant power converters, which include the effective generation of RPCs as a high-switching converter that can serve as a solution to electromagnetic interference (EMI) and switching losses difficulties that surface by using the PWM converters. Furthermore, the resonant converter classi- fications depending on several points of views as well as controlled techniques (either varied or fixed frequency techniques) were elucidated. The controlled techniques and several application areas of resonant converters have been explained.

The variable and fixed frequency controls are used to cross-check the LLC converter output voltage. The load is changed when using 50% of the full load in the entire control techniques; the obtained results affirmed the significant stable response of both controllers with little overshoot voltage. Nevertheless, the variable frequency control gives a significant outcome based on the resonant tank wave- forms in relative to the stable frequency control when changing the load.

Author details

Mohammed Salem[1]* and Khalid Yahya[2]

1 School of Electrical and Electronic Engineering, Universiti Sains Malaysia, Nibong Tebal, Malaysia

2 Kocaeli University, İzmit, Kocaeli, Turkey

*Address all correspondence to: salemm@usm.my

References

[1] Outeiro MT, Visintini R, Buja G. Considerations in designing power supplies for particle accelerators. In: Industrial Electronics Society, IECON 2013-39th Annual Conference of the IEEE 2013 Nov 10. IEEE; 2013. pp. 7076-7081. DOI: 10.1109/ IECON.2013.6700307

[2] Salem M, Jusoh A, Idris NR. Implementing buck converter for battery charger using soft switching techniques. In: Power Engineering and Optimization Conference (PEOCO), 2013 IEEE 7th International 2013 Jun 3. IEEE; 2013. pp. 188-192. DOI: 10.1109/ PEOCO.2013.6564540

[3] Mostaghimi O, Wright N, Horsfall A. Design and performance evaluation of SiC based DC-DC converters for PV applications. In: Energy Conversion Congress and Exposition (ECCE), 2012 IEEE 2012 Sep 15. IEEE; 2012. pp. 3956-3963. DOI: 10.1109/ ECCE.2012.6342163

[4] Ching TW, Chan KU. Review of soft- switching techniques for high- frequency switched-mode power converters. In: Vehicle Power and Propulsion Conference, 2008. VPPC'08. IEEE 2008 Sep 3. IEEE; 2008. pp. 1-6. DOI: 10.1109/VPPC.2008.4677473

[5] Shafiei N, Pahlevaninezhad M, Farzanehfard H, Bakhshai A, Jain P. Analysis of a fifth-order resonant converter for high-voltage DC power supplies. IEEE Transactions on Power Electronics. 2013;28(1):85-100. DOI: 10.1109/TPEL.2012.2200301

[6] Outeiro MT, Buja G, Carvalho A. Resonant converters for electric equipment power supply. In: Industrial Electronics Society, IECON 2014-40th Annual Conference of the IEEE 2014 Oct 29. IEEE; 2014. pp. 5065-5071. DOI: 10.1109/IECON.2014.7049270

[7] Salem M, Jusoh A, Idris NR, Sutikno T, Buswig YM. Phase-shifted series resonant converter with zero voltage switching turn-on and variable frequency control. International Journal of Power Electronics and Drive Systems (IJPEDS). 2017;8(3):1184-1192. DOI: 10.11591

[8] Jin K, Ruan X. Hybrid full-bridge three-level LLC resonant converter-a novel DC-DC converter suitable for fuel cell power system. In: Power Electronics Specialists Conference, 2005. PESC'05. IEEE 36th 2005 Jun 16. IEEE; 2015. pp. 361-367. DOI: 10.1109/ PESC.2005.1581649

[9] Salem M, Jusoh A, Rumzi N, Idris N, Alhamrouni I. Steady state and generalized state space averaging analysis of the series resonant converter. In: 3rd IET Inter-national Conference on Clean Energy and Technology (CEAT). 2014. pp. 1-5. DOI: 10.1049/ cp.2014.1488

[10] Alhamroun I, Salem M, Jusoh A, Idris NR, Ismail B, Albatsh FM. Comparison of two and four switches inverter feeding series resonant converter. In: Energy Conversion (CENCON), 2017 IEEE Conference on 2017 Oct 30. IEEE; 2017. pp. 334-338. DOI: 10.1109/ CENCON.2017.8262508

[11] Rathore AK, Bhat AK, Oruganti R. A comparison of soft-switched DC-DC converters for fuel cell to utility interface application. IEEJ Transactions on Industry Applications. 2008;128(4): 450-458. DOI: 10.1541/ieejias.128.450

[12] Salem M, Jusoh A, Idris NR, Alhamrouni I. Comparison of LCL resonant converter with fixed frequency, and variable frequency controllers. In: Energy Conversion (CENCON), 2017 IEEE Conference on 2017 Oct 30. IEEE; 2017. pp. 84-89. DOI: 10.1109/ CENCON.2017.8262463

[13] Yang Y, Huang D, Lee FC, Li Q. Analysis and reduction of common mode EMI noise for resonant converters. In: Applied Power Electronics Conference and Exposition (APEC), 2014 Twenty-Ninth Annual IEEE 2014 Mar 16. IEEE; 2014. pp. 566-571. DOI: 10.1109/ APEC.2014.6803365

[14] Outeiro MT, Buja G, Czarkowski D. Resonant power converters: An overview with multiple elements in the resonant tank network. IEEE Industrial Electronics Magazine. 2016;10(2):21-45. DOI: 10.1109/ MIE.2016.2549981

[15] Polleri A, Anwari M. Modeling and simulation of paralleled series-loaded- resonant converter. In: Second Asia International Conference on Modelling & Simulation. 2008. p. 974. DOI: 10.1109/AMS.2008.86

[16] Salem M, Jusoh A, Idris NR, Das HS, Alhamrouni I. Resonant power converters with respect to passive storage (LC) elements and control techniques–An overview. Renewable and Sustainable Energy Reviews. 2018; 91:504-520. DOI: 10.1016/j. rser.2018.04.020

[17] Chuang YC, Ke YL, Chuang HS, Chen HK. Implementation and analysis of an improved series-loaded resonant DC-DC converter operating above resonance for battery chargers. IEEE Transactions on Industry Applications. 2009;45(3):1052-1059. DOI: 10.1109/ TIA.2009.2018946

[18] Ivensky G, Bronshtein S, Abramovitz A. Approximate analysis of resonant LLC DC-DC converter. IEEE Transactions on Power Electronics. 2011;26(11):3274-3284. DOI: 10.1109/ TPEL.2011.2142009

[19] Salem M, Jusoh A, Idris NR, Alhamrouni I. A review of an inductive power transfer system for EV battery charger. European Journal of Scientific Research. 2015;134:41-56

[20] Tan X, Ruan X. Equivalence relations of resonant tanks: A new perspective for selection and design of resonant converters. IEEE Transactions on Industrial Electronics. 2016;63(4): 2111-2123. DOI: 10.1109/ TIE.2015.2506151

[21] Shafiei N, Farzanehfard H. Steady state analysis of LCLC resonant converter with capacitive output filter. In: Power and Energy (PECon), 2010 IEEE International Conference on 2010 Nov 29. IEEE; 2010. pp. 807-812. DOI: 10.1109/PECON.2010.5697690

[22] Lin BR, Hou BR. Analysis and implementation of a zero-voltage switching pulse-width modulation resonant converter. IET Power Electronics. 2014;7(1, 1):148-156. DOI: 10.1049/iet-pel.2013.0134

[23] Salem M, Jusoh A, Idris NR, Sutikno T, Abid I. ZVS full bridge series resonant boost converter with series- connected transformer. International Journal of Power Electronics and Drive Systems (IJPEDS). 2017;8(2):812-825. DOI: 10.11591

[24] Salem M, Jusoh A, Idris NR, Tan CW, Alhamrouni I. Phase-shifted series resonant DC-DC converter for wide load variations using variable frequency control. In: Energy Conversion (CENCON), 2017 IEEE Conference on 2017 Oct 30. IEEE; 2017. pp. 329-333. DOI: 10.1109/CENCON.2017.8262507

[25] Lu Y, Cheng KW, Ho SL, Pan JF. Passivity-based control of a phase- shifted resonant converter. IEE Proceedings-Electric Power Applications. 2005;152(6):2005, 1509-2015. DOI: 10.1049/ip-epa:20045261

[26] Lu Y, Cheng KE, Ho SL. Quasi current mode control for the phase- shifted series resonant converter. IEEE Transactions on Power Electronics. 2008;23(1):353-358. DOI: 10.1109/ TPEL.2007.911846

[27] Salem M, Jusoh A, Idris NR, Alhamrouni I. Extension of zero voltage switching range for series resonant converter. In: Energy Conversion (CENCON), 2015 IEEE Conference on 2015 Oct 19. IEEE; 2015. pp. 171-175. DOI: 10.1109/CENCON.2015.7409534

[28] Musavi F, Craciun M, Gautam DS, Eberle W, Dunford WG. An LLC resonant DC-DC converter for wide output voltage range battery charging applications. IEEE Transactions on Power Electronics. 2013;28(12): 5437-5445. DOI: 10.1109/ TPEL.2013.2241792

[29] Mousavi SM, Beiranvand R, Goodarzi S, Mohamadian M. Designing A 48 V to 24 V DC-DC converter for vehicle application using a resonant switched capacitor converter topology. In: Power Electronics, Drives Systems & Technologies Conference (PEDSTC), 2015 6th 2015 Feb 3. IEEE; 2015. pp. 263-268. DOI: 10.1109/ PEDSTC.2015.7093285

[30] Bojarski M, Asa E, Outeiro MT, Czarkowski D. Control and analysis of multi-level type multi-phase resonant converter for wireless EV charging. In: Industrial Electronics Society, IECON 2015-41st Annual Conference of the IEEE 2015 Nov 9. IEEE; 2015. pp. 5008-5013. DOI: 10.1109/ IECON.2015.7392886

[31] Outeiro MT, Carvalho A. Methodology of designing power converters for fuel cell based systems: A resonant approach. In: New Developments in Renewable Energy. Rijeka: In Tech; 2013. DOI: 10.5772/ 54674

[32] Outeiro MT, Carvalho A. Design, implementation and experimental validation of a DC-DC resonant converter for PEM fuel cell applications. In: Industrial Electronics Society, IECON 2013-39th Annual Conference of the IEEE 2013 Nov 10. IEEE; 2013. pp. 619-624. DOI: 10.1109/ IECON.2013.6699206

[33] Rezaei MA, Lee KJ, Huang AQ. A high-efficiency flyback micro- inverter with a new adaptive snubber for photovoltaic applications. IEEE Transactions on Power Electronics. 2016;31(1, 1):318-327. DOI: 10.1109/ TPEL.2015.2407405

[34] Beres RN, Wang X, Blaabjerg F, Liserre M, Bak CL. Optimal design of high-order passive-damped filters for grid-connected applications. IEEE Transactions on Power Electronics. 2016;31(3):2083-2098. DOI: 10.1109/ TPEL.2015.2441299

[35] Mishima T, Nakagawa Y, Nakaoka M. A bridgeless BHB ZVS-PWM AC-AC converter for high-frequency induction heating applications. IEEE Transactions on Industry Applications. 2015;51(4): 3304-3315. DOI: 10.1109/TIA.2015.24 09177

8

Power Device Loss Analysis of a High-Voltage High-Power Dual Active Bridge DC-DC Converter

Thaiyal Naayagi Ramasamy

Abstract

The insulated-gate bipolar transistor (IGBT) offers low conduction loss and improved performance and, hence, is a potential candidate for high-current and high-voltage power electronic applications. This chapter presents the power loss estimation of IGBTs as employed in a high-voltage high-power dual active bridge (DAB) DC-DC converter. The mathematical models of the device currents are derived, and the power loss prediction is clearly explained using the mathematical models. There are many parameters to consider when selecting an appropriate power device for a given application. This chapter highlights the step-by-step procedure for selecting suitable IGBTs for a 20 kW, 540/125 V, 20 kHz DAB converter designed for aerospace energy storage systems. Experimental results are given to demonstrate the device performance at 540 V, 80 A operation of high-voltage IGBTs and 125 V, 300 A operation of low-voltage IGBTs and thus validate the selection procedure presented.

Keywords: power loss prediction, HV-side IGBTs, LV-side IGBTs, zero voltage switching, zero current switching, dual active bridge, DC-DC converter

Introduction

The dual active bridge (DAB) [1, 2] is a bidirectional DC-DC converter that has a small transformer between the two active bridges. This topology has received significant attention from researchers for high-power applications for decades.

Compared with a single-phase DAB, a three-phase DAB of similar rating offers the benefits of smaller passive components and improved magnetic utilization [3]. Acronyms used in this chapter are listed in **Table 1**. A high-voltage SiC IGBT is used to achieve optimum design of a DAB converter based on the consideration of the device and transformer characteristics in [4]. However, the high dv/dt of the SiC IGBT switches [5–7] causes a huge spike and ringing in the currents during hard switching at high voltage levels.

An optimization algorithm for non-active power loss minimization in the DAB DC-DC converter is discussed in [8] to improve the converter efficiency. Ref. [9] presents an advanced control strategy to reduce DAB converter losses. A survey of material-level developments in the key components of PWM power converters and the potential for improving system power density using advanced components is presented in [10]. Ref. [11, 12] indicate that a reduction of the switching loss by tenfold is achievable with the use of such advanced devices. Although wide bandgap devices have merits over Si devices, most have not been commercialized due to various reasons, such as the difficulty of mass producing large-diameter SiC wafers, the difficulty of controlling the impurity level of these devices, and high cost. Several parameters must be considered to select the correct power semiconductor device for any power electronic circuit. Device output characteristics, thermal characteristics, switching characteristics, blocking voltage, leakage current, and safe operating area (SOA) are all important factors with respect to reliability. It is important to note that at any point of time during device operation, the maximum ratings specified in the device datasheet should not be exceeded [13–15]. The gate charge characteristics and gate driver requirements of a power device and the associated driver circuit loss are also important factors to be considered when selecting an appropriate power device. The focus of this chapter is the estimation of the conduction power loss and the switching power loss. The key parameters and the device characteristics portrayed in the datasheets are discussed in [16].

Table 1. *List of acronyms and their abbreviations.*

Acronyms	Abbreviations
SiC	Silicon carbide
PWM	Pulse width modulation
HV	High voltage
LV	Low voltage
DAB	Dual active bridge
IGBT	Insulated-gate bipolar transistor
MOSFET	Metal-oxide-semiconductor field-effect transistor
DC	Direct current
ZVS	Zero voltage switching
ESR	Effective series resistance
ZCS	Zero current switching

The DAB DC-DC converter is shown in **Figure 1**. The converter has two H-bridge circuits interfaced via a high-frequency transformer. To correctly choose the power device for the DAB, power loss calculation is performed for all power devices under the worst-case operating condition of the converter.

The performances of various metal-oxide-semiconductor field-effect transistors (MOSFETs) and IGBTs produced by commercial manufacturers are compared to determine the appropriate power semiconductor device for the high-voltage (HV) and low-voltage (LV) sides of the converter [17–21]. For mission-critical applications such as aerospace systems, voltage transients are inevitable and are taken into consideration when selecting the voltage ratings of devices: 1200 V IGBTs are chosen for the HV side, and 600 V IGBTs are chosen for the LV side [22]. In addition to the device voltage and current handling capabilities and the power density, the anti-parallel diode's on-state voltage and soft recovery characteristics should be taken into consideration. An analysis of MOSFET usage for the low-voltage side of the converter is presented in [23]. Due to the low voltage on the secondary side, the devices should be able to handle high currents. Hence, three or more MOSFETs should be connected in parallel, which makes the circuit very complex. Consequently, IGBT technology [24–26] has been chosen due to its high-current and high-voltage handling capabilities. The IGBT power modules have the benefit of reduced internal inductance and improved heat dissipation and are easy to connect. This chapter discusses the selection procedure of appropriate power devices for a high-power DAB DC-DC converter design. The mathematical model of the device current equations is presented. Researchers at an early stage of their research will find this information valuable as a reference for the design and development of converter prototypes.

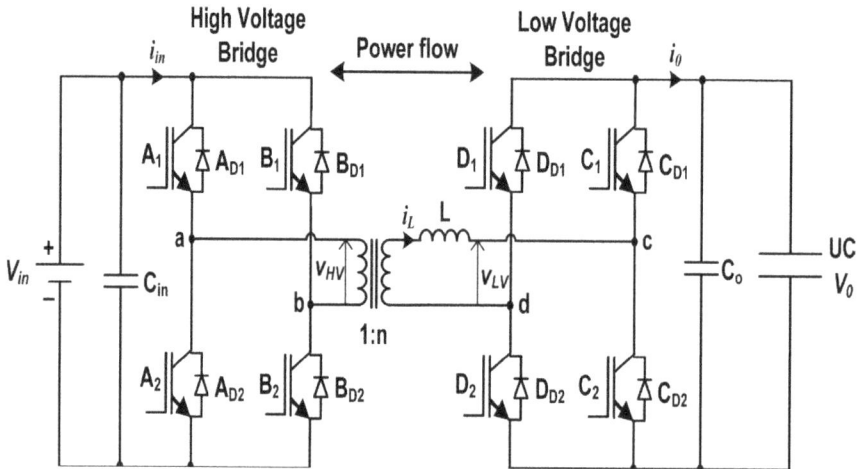

Figure 1. *Dual active bridge DC-DC converter.*

This chapter significantly extends the experimental results given in [16]. Specifically, the experimental results for high-voltage IGBTs with the snubber capacitor and for low-voltage IGBTs with and without the snubber capacitors are included. In addition, the experimental measurements of power losses of the HV- and LV-side IGBTs are presented. The remainder of this chapter is organized as follows. The power loss estimation of HV- IGBTs of the converter from various manufacturers is presented in Section 2. Section 3 discusses the power loss calculation of LV-side IGBTs and the loss comparison of IGBTs with similar ratings from various leading manufacturers. The experimental results are presented in Section 4, and the conclusions are given in Section 5.

Power loss estimation of HV-side devices of the DAB DC-DC converter

To select the correct device, a power loss calculation should be performed for the worst-case scenario of the converter for the chosen application. In this work, we consider the design of a DAB converter for an aerospace energy storage system whose HV side is connected to the DC link of an aircraft and its LV side is connected to an ultracapacitor-based energy storage system. The working range for the ultracapacitor is assumed to be 2:1. The worst-case operating condition of the converter is $V_{HV} = 540$ V, $V_{LV} = 62.5$ V, inductor RMS current $I_{RMS} = 427$ A, peak inductor current $I_p = 640$ A, average ultracapacitor current $I_o = 320$ A, power throughput $P_o = 20$ kW, and $d = 0.5$. The main component values of the converter are determined by applying the worst-case operating condition to the equations derived based on the mathematical model of the DAB converter, which are presented in [16, 27].

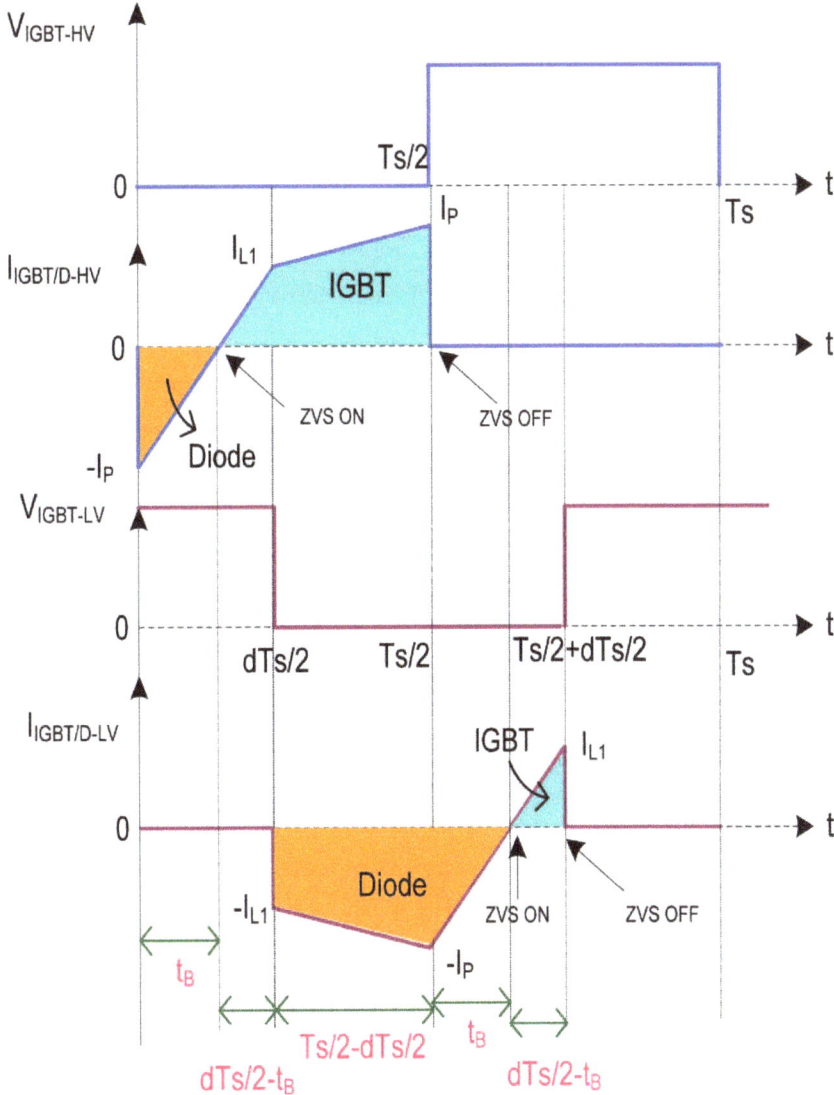

Figure 2. *Theoretical waveforms of devices on the HV side and LV side of the DAB DC-DC converter. (Device conduction intervals are marked in pink).*

The loss calculation is performed for the devices provided by various leading manufacturers for a 20 kHz switching frequency. The theoretical waveforms of the converter and the device currents are illustrated in **Figure 2**. The terms and symbols used in the analysis are listed in **Table 2**.

Piecewise linearity is assumed for the device current waveforms. Hence, the average current flow through the IGBT on the HV side of the converter during the on-time is given by

$$I_T = \frac{\frac{1}{2} \times I_{L1} \times \left(\frac{dT_S}{2} - t_B\right) + \frac{1}{2} \times (I_{L1} + I_P) \times \left(\frac{T_S}{2} - \frac{dT_S}{2}\right)}{\frac{T_S}{2} - t_B} \qquad (1)$$

Table 2. *List of symbols/terms used in analysis and their description.*

Symbols/terms	Description
d	Duty cycle
T_s	Switching time period
I_T	Transistor current
I_P	Peak inductor current
I_{L1}	First inductor current peak
t_B	Time interval from I_P to zero current
$V_{CE\,(sat)}$	Transistor forward voltage drop
P_{CondT}	Transistor conduction loss
r_{CE}	Collector to emitter resistance
V_{CEO}	Collector to emitter voltage with gate open
I_D	Diode current
P_{CondD}	Diode conduction loss
V_F	Diode forward voltage drop
P_C	Total conduction loss
P_{ON}	Transistor turn-on power loss
E_{ON}	Transistor turn-on energy loss
f	Switching frequency
P_{OFF}	Transistor turn-off power loss
E_{OFF}	Transistor turn-off energy loss
V_{CE}	Collector to emitter voltage
I_C	Collector current
t_r	Rise-time
t_f	Fall-time
P_{SW}	Total switching loss
$R_{th(c-s)}$	Case to sink thermal resistance
$R_{th(j-c)}$	Junction to case thermal resistance
P_T	Total power loss
T_{jIGBT}	Junction temperature of IGBT
T_{jDiode}	Junction temperature of diode

The IGBT has a constant voltage drop during the on-state. Hence, the conduction loss can be calculated using the average IGBT current and the duty cycle, which is given as

$$P_{CondT} = V_{CE(sat)} \times I_T \times d \tag{2}$$

The duty cycle is the ratio of the on-period of the transistor to the switching period. The datasheet specifies the forward voltage drop of the transistor ($V_{CE(sat)}$) and the anti-parallel diode (V_F) with respect to the main terminals of the modules, which includes the voltage drops across the terminals. Due to the high power densities of the devices, terminal losses cannot be neglected compared with the semiconductor losses. Hence, it is important to specify the voltage drop at the chip level and across the terminals (r_{CE}') separately. The voltage drop across the terminals is given by

$$V_{CE(sat)} = V_{CEO} + (r_{CE} \times I_T) \tag{3}$$

The average diode current on the HV side of the converter during the on-state, as shown in **Figure 2**, is given as

$$I_D = \frac{1}{2} \times I_P \tag{4}$$

In Eq. (4), the diode conduction interval and the base interval are the same on the HV side; hence, the diode current is computed regardless of time. Assuming a constant voltage drop, the conduction losses of the diode are estimated as

$$P_{CondD} = V_F \times I_D \times d \tag{5}$$

The DAB converter has four transistors and four antiparallel diodes on the HV side. Hence, the total conduction loss P_C of the semiconductor devices on the HV side of the converter is given as

$$P_C = 4 \times (P_{CondT} + P_{CondD}) \tag{6}$$

Switching losses are calculated from the turn-on (E_{ON}) and turn-off (E_{OFF}) energy loss curves, which are usually given as a function of the collector current at the switching instants of an IGBT, as given in the manufacturer datasheet. In the DAB converter configuration under zero voltage switching (ZVS), the diode always turns off with zero current, thus eliminating the diode reverse recovery losses.

Therefore, the power loss equations for the turn-on and turn-off instants are

$$P_{ON} = E_{ON} \times f \tag{7}$$

and

$$P_{OFF} = E_{OFF} \times f \tag{8}$$

P_{ON} is the power loss during the turn-on instant, P_{OFF} is the power loss during the turn-off instant, and f is the switching frequency. In some manufacturers' datasheets, the energy loss curves for turn-on and turn-off may not be available. For those cases, it is necessary to consider the simplified voltage and current waveforms during the switching process. Switching losses are predominant during the risetime and fall-time periods. Hence, the loss equations are approximated during the turn-on and turn-off instants, which are given as

$$P_{ON} = \frac{1}{2}V_{CE} \times I_C \times t_r \times f \tag{9}$$

and

$$P_{OFF} = \frac{1}{2}V_{CE} \times I_C \times t_f \times f \tag{10}$$

In DAB converter operation during ZVS, the anti-parallel diode current is always transferred to the IGBTs. Hence, the turn-on switching losses can be neglected.

Therefore, the total switching losses (P_{SW}) are approximated as

$$P_{SW} = 4 \times [E_{OFF} \times f] \tag{11}$$

The junction temperature of the devices is calculated as

$$T_{jIGBT} = T_s + \left(P_T \times R_{th(c-s)}\right) + \left(P_{IGBT} \times R_{th(j-c)}\right) \tag{12}$$

and

$$T_{jDiode} = T_s + \left(P_T \times R_{th(c-s)}\right) + \left(P_{Diode} \times R_{th(j-c)}\right) \tag{13}$$

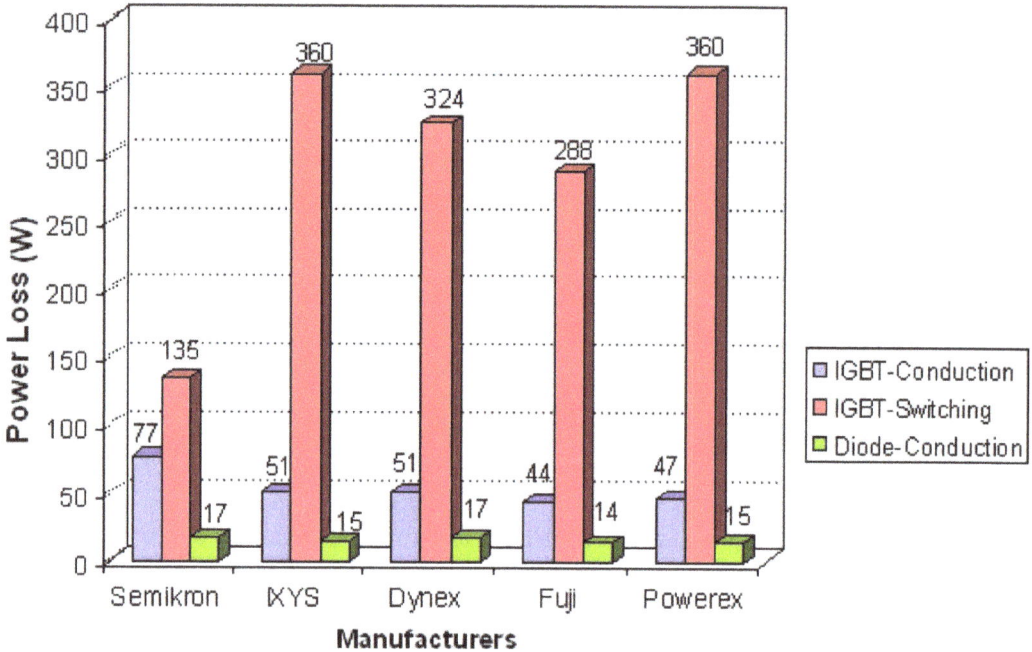

Figure 3. *HV-side IGBT and anti-parallel diode power loss comparison for various manufacturers.*

where T_S is the heat sink temperature and P_T is the total power loss of the IGBT module. The case to sink $R_{th(c-s)}$ and junction to case $R_{th(j-c)}$ thermal resistances can be obtained from the device datasheet. The total power losses of the IGBTs and the antiparallel diode of the HV side of the DAB converter are compared for devices with ratings similar to 1200 V, 300 A from five different manufacturers:

- SKM300GB125D from Semikron, which has a rating of 1200 V, 300 A at $T_{case} = 25°C$;

- MII300-12A4 from IXYS, which has a rating of 1200 V, 330 A at T_{case} = 25°C;

- DIM200WHS12-A000 from Dynex Semiconductor, which has a rating of 1200 V, 200 A at T_{case} = 80°C;

- 2MB1300U4H-120 from Fuji Semiconductor, which has a rating of 1200 V, 400 A at T_{case} = 25°C; and

- CM300DY-24A from Powerex, which has a rating of 1200 V, 300 A at T_{case} = 25°C.

The power losses of IGBTs and anti-parallel diodes are predicted for the worst- case operating condition and are depicted in **Figure 3**. Based on a comparison of the losses of all devices, the ultra-fast SKM300GB125D 1200 V, 300 A phase leg IGBT modules from Semikron [28] are selected for the HV side of the converter due to their low loss. As shown in **Figure 3**, the switching power losses of the IGBTs are significant and are a major contributor to the overall power loss due to the high switching frequency and high turn-off current.

Power loss estimation of LV-side devices of the DAB DC-DC converter

The power loss of LV-side devices is predicted using a procedure similar to that discussed in Section 2. The main difference is that the shape of the LV-side device currents differs from those of the HV-side devices due to the phase shift introduced between the HV- and LV-side bridges as shown in **Figure 2**. The transistor and diode current equations during the on-state are

$$I_T = \frac{\frac{1}{2} \times I_{L1} \times \left(\frac{dT_S}{2} - t_B\right)}{\left(\frac{dT_S}{2} - t_B\right)} \tag{14}$$

and

$$I_D = \frac{\frac{1}{2} \times (I_P \times t_B) + \frac{1}{2} \times (I_{L1} + I_P) \times \left(\frac{T_S}{2} - \frac{dT_S}{2}\right)}{\left(\frac{T_S}{2} - \frac{dT_S}{2} + t_B\right)} \tag{15}$$

Similar to the HV side, the on-state and switching power losses of the IGBTs and the antiparallel diode of the LV side of the DAB converter are compared for devices with ratings similar to 600 V, 700 A at T_{case} = 25°C from four leading manufacturers:

- SKM600GB066D from Semikron, which has a rating of 600 V, 760 A;

- CM600DY-12NF from Mitsubishi, which has a rating of 600 V, 600 A;

- 2MB1600U2E-060 from Fuji Semiconductor, which as a rating of 650 V, 600 A; and

- CM600DY-12NF from Powerex, which has a rating of 600 V, 600 A.

Figure 4 portrays the power loss comparison of an LV IGBT and antiparallel diode for each module. As shown in **Figure 4**, all IGBT modules exhibit similar performance. During the forward conduction mode (charging mode), the diodes perform the rectification function. Hence, the conduction losses of the diodes are predominant. When the power flow reverses, the loss values are exchanged between the diode and IGBT. The phase leg modules from Semikron have lower losses than the other devices. In addition, the Semikron module has a maximum junction temperature of 175°C, whereas the other modules have a maximum junction temperature of 150°C. Hence, the Semikron SKM600GB066D high-temperature phase leg IGBT modules [29], which have a rating of 600 V, 760 A, were chosen for the LV side of the converter.

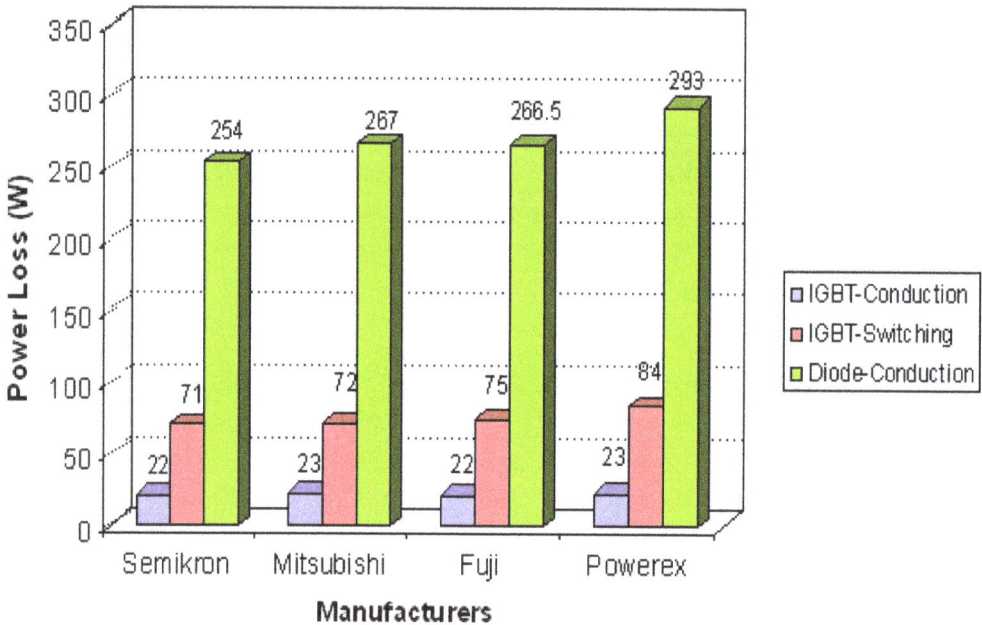

Figure 4. *LV-side IGBT and antiparallel diode power loss comparison from various manufacturers.*

Experimental results

Experiments were performed for the HV- and LV-side devices of the converter using a suitable reactive load. A 39 μH air-core inductor was used for HV bridge testing by phase shifting the two half-bridge legs of the IGBTs, and the devices are subjected to maximum currents. Because the load was purely reactive, the current drawn from the source was used to meet the device conduction losses, switching losses, and the losses due to the passive components. A photograph of the DAB converter prototype is shown in **Figure 5**. Experimental results are shown in **Figures 6** and **7** for the HV-side IGBTs of the DAB converter for 540 V, 80 A operation with and without the snubber capacitor.

A 47 nF snubber capacitor was used on the HV side to minimize the turn-off losses and improve the switching transient of the HV-side IGBTs. A direct mount snubber capacitor type was used with a reduced effective series resistance (ESR) and with a low effective series inductance (ESL). As evident from the waveforms shown in **Figures 6** and **7**, the diode begins conduction before the transistor, which ensures ZVS turn-on at peak current and ZVS/ZCS during turn-off. Similarly, the IGBT has ZVS/ZCS turn-on and ZVS turn-off at peak current.

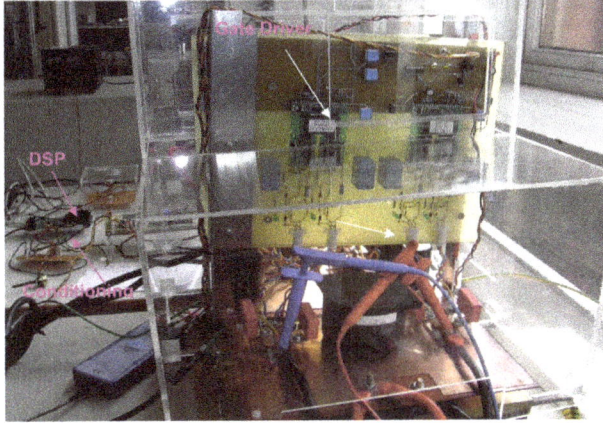

(a)

(b)

Figure 5. *Photographs of the DAB converter prototype: (a) front view and (b) back view.*

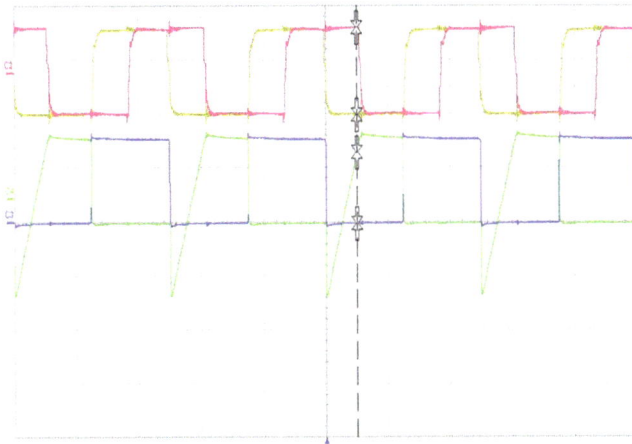

Figure 6. *Experimental results for the HV-side converter. V_{in} = 540 V, L = 39 μH, V_{Lrms} = 363 V, I_{Lrms} = 65.6 A, f_s = 20 kHz, I_{OFF} = 80 A. Channel 1 (yellow)—gate pulse of transistor A_2, 10 V/div. Channel 2 (pink)—gate pulse of transistor B_2, 10 V/div. Channel 3 (blue)—IGBT A_1 voltage, 350 V/div. Channel 4 (green)—IGBT A_1 current, 50A/div. Time scale: 20 μs/div.*

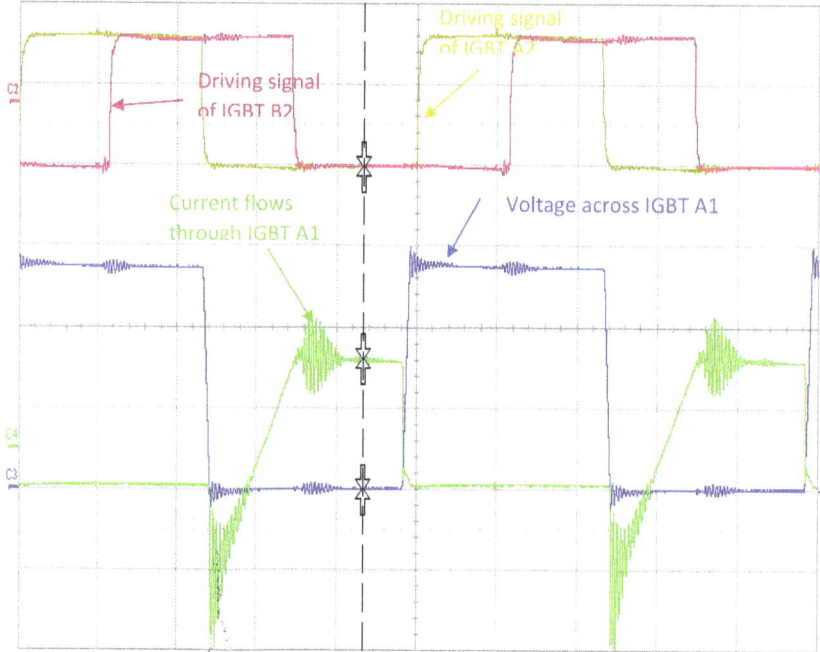

Figure 7. *Experimental results for the HV-side converter with a 47 nF snubber. V_{in} = 540 V, I_{in} = 1.318 A, L = 39 μH, f_s = 20 kHz, V_{Lrms} = 363 V, I_{Lrms} = 65.6 A, I_{OFF} = 80 A, R_G = 3 Ω, C_s = 47 nF. Time scale: 10 μs/div. Channel 1 (yellow)—gate pulse of transistor A2, 10 V/div. Channel 2 (pink)—gate pulse of transistor B2, 10 V/div. Channel 3 (blue)—IGBT A1 voltage, 200 V/div. Channel 4 (green)—IGBT A1 current, 50A/div.*

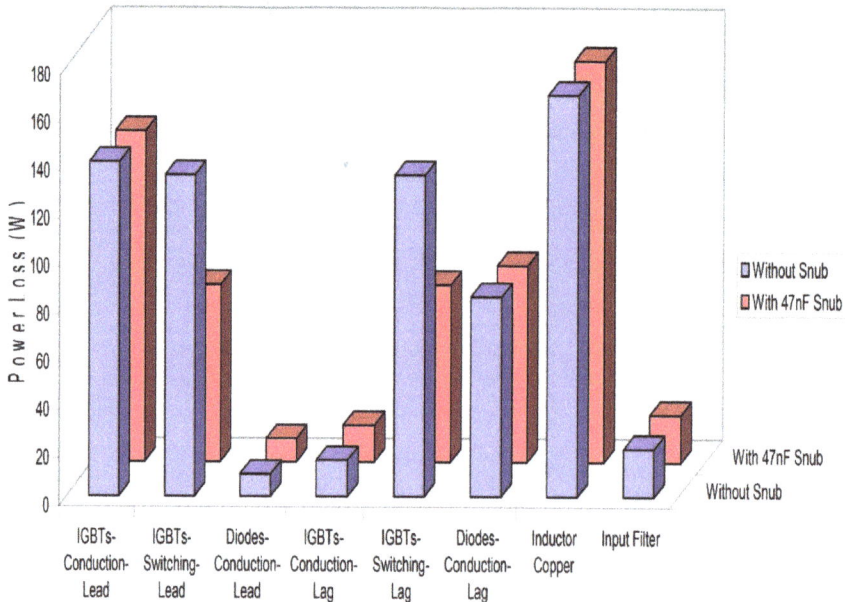

Figure 8. *Experimental power loss breakdown of the HV-side converter with and without the snubber capacitor. V_{in} = 540 V, L = 39 μH, f_s = 20 kHz, I_{OFF} = 80 A, V_{Lrms} = 363 V, I_{Lrms} = 65.6 A, R_G = 3 Ω.*

The power loss measured during the experiments for the HV bridge converter at 16.6 kVA operation is plotted in **Figure 8**. Two cases are considered: in the first, a 47 nF snubber capacitor is connected across each of the four IGBTs of the HV-side bridge; in the second, no snubber capacitor is used. The power losses of the leading-phase and lagging-phase leg devices are displayed separately in **Figure 8** due to different conduction intervals of the devices. Power losses in the cables, busbar, and connections are neglected. Due to the ZCS turn-off of the diodes, their reverse recovery losses are neglected; due to ZVS/ZCS turn-on of the IGBTs, their turn-on losses are omitted. As shown in **Figure 8**, the air-core inductor copper losses are high although it has only 11 turns. Diode and IGBT conduction losses, however, remain approximately constant with or without the snubber capacitors. As evident in **Figure 8**, there is a considerable reduction in turn-off losses when the snubber capacitor is used.

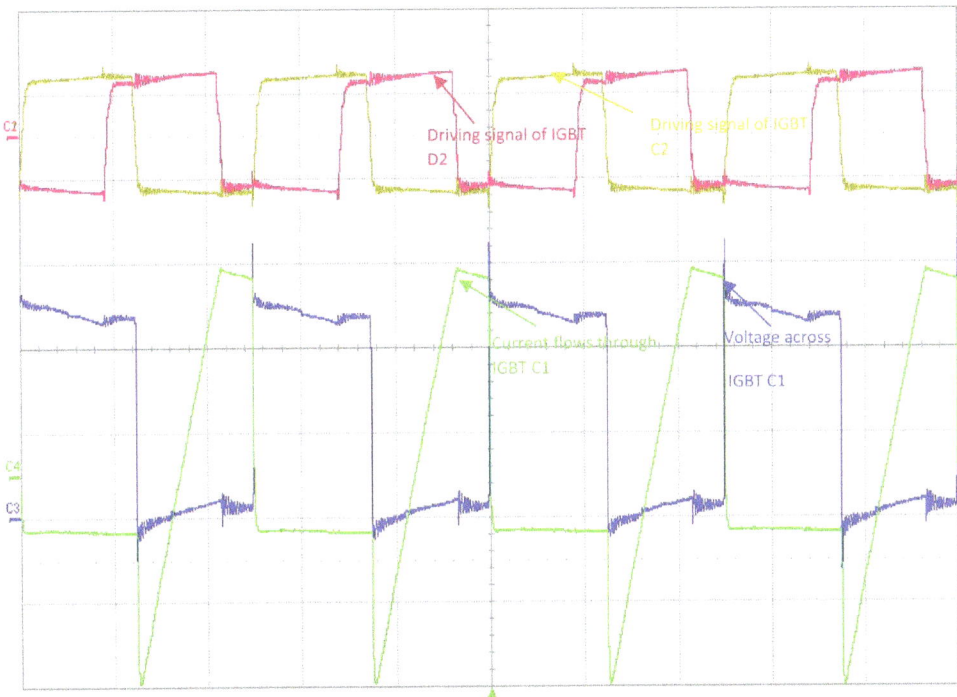

Figure 9. *Experimental results for the LV-side converter without the snubber. V_{in} = 125 V, I_{in} = 10.51 A, L = 4.17 μH, V_{Lrms} = 106 V, I_{Lrms} = 202 A, f_s = 20 kHz, I_{OFF} = 300 A, R_G = 2.5 Ω. Time scale: 20 μs/div. Channel 1 (yellow)— gate pulse of transistor C2, 10 V/div. Channel 2 (pink)—gate pulse of transistor D2, 10 V/div. Channel 3 (blue)—IGBT C1 voltage, 100 V/div. Channel 4 (green)—IGBT C1 current, 100A/div.*

The LV-side devices of the converter were tested, and the LV-side IGBTs were subjected to a peak current of 300 A at 125 V. The two legs of the LV-side bridge were connected through an air core inductor with a value of 4.17 μH. An external electrolytic capacitor bank of 19.8 mF capacitance (comprising three 6.6 mF capacitor banks) was added to the DC supply to smooth the input ripple current. **Figure 9** shows the current and voltage waveforms of the leading-leg IGBT and the driving signals of the IGBTs, D_2 and C_2. A CWT15 Rogowski current probe with a sensitivity of 2 mV/A was used to measure the device currents. The measured current fall-time is 205 ns for the initial current fall time and 620 ns for the tail current fall duration. High-frequency ringing is observed in the device waveforms when the snubbers are introduced across the devices. This may be resonance due to parasitic inductances of the module (with values of 15–20 nH), the busbars, and the snubber capacitor connections.

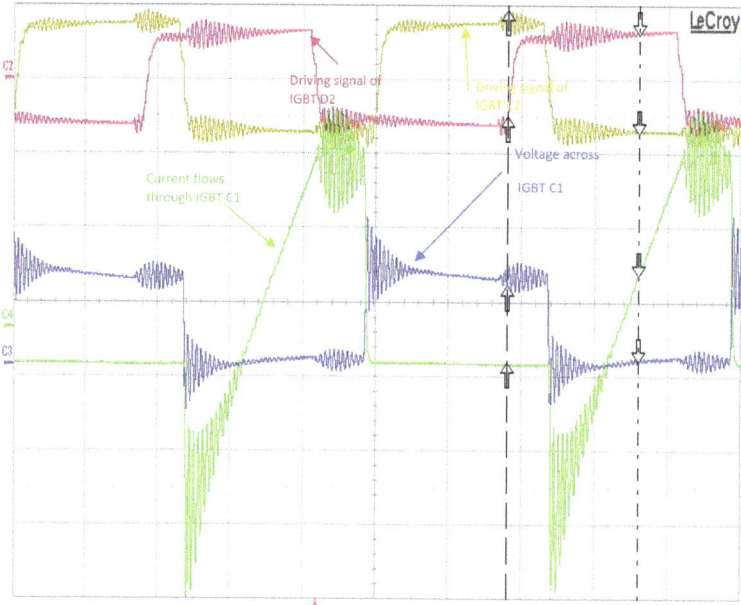

Figure 10. *Experimental waveforms of the LV-side converter with a 100 nF snubber. $V_{in} = 125$ V, $I_{in} = 9.29$ A, L = 4.17 µH, $V_{Lrms} = 106$ V, $I_{Lrms} = 202$ A, $f_s = 20$ kHz, $I_{OFF} = 300$ A, $R_G = 2.5$ Ω. Time scale: 10 µs/div. Channel 1 (yellow)— gate pulse of transistor C2, 10 V/div. Channel 2 (pink)—gate pulse of transistor D2, 10 V/div. Channel 3 (blue)—IGBT C1 voltage, 100 V/div. Channel 4 (green)—IGBT C1 current, 100A/div.*

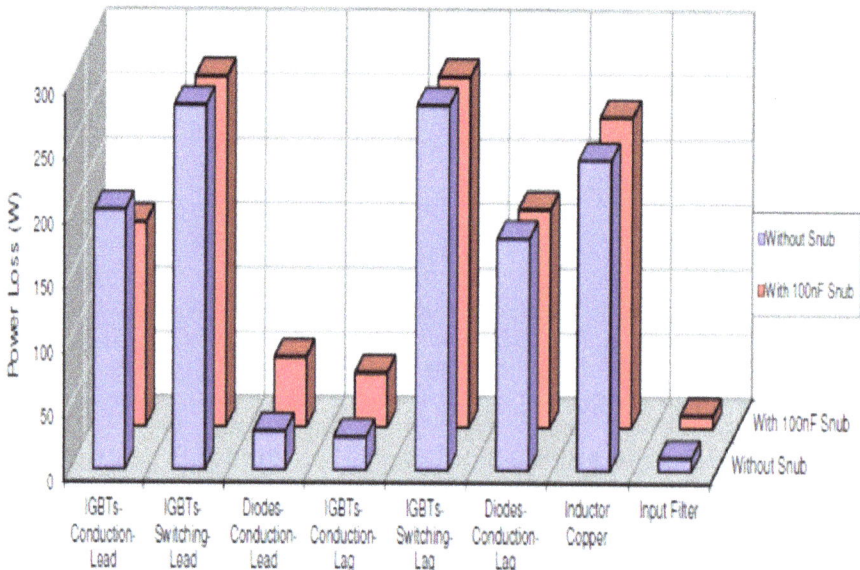

Figure 11. *Experimental measurement of power loss on the LV-side converter with and without the snubber. V_{in} = 125 V, L = 4.17 µH, f_s = 20 kHz, I_{OFF} = 300 A, V_{Lrms} = 106 V, I_{Lrms} = 202 A, R_G = 2.5 Ω.*

The input power drawn from the source using 100 nF snubber capacitors is lower than those observed for other snubbers. Moreover, a significant reduction in *dv/dt* during the switching transients is observed. Hence, 100 nF snubbers are chosen for the LV-side IGBTs. **Figure 10** shows the experimental waveforms for the LV-side IGBT C_1 with a 100 nF snubber capacitor. **Figure 11** illustrates the power loss breakdown for the LV-side bridge converter at 19.2 kVA operation with and without the snubber capacitors, as determined from the measurements. The IGBT switching losses are dominant. Power losses due to cables, busbars, and connections are not included. When 100 nF snubber capacitors are connected across the IGBTs, no significant reduction in switching losses occurs. The experimental power loss using the 100 nF snubber is 135 W, and *dv/dt* during the switching transient is reduced significantly, thereby minimizing the device stresses.

Comparisons of the theoretical and practical power losses of the HV- and LV-side devices are shown in **Figures 12** and **13**, respectively, for the leading leg devices to validate the mathematical models describing the device current equations. A close correlation between the calculated and experimental power loss values is observed, which demonstrates the effectiveness of the mathematical models presented.

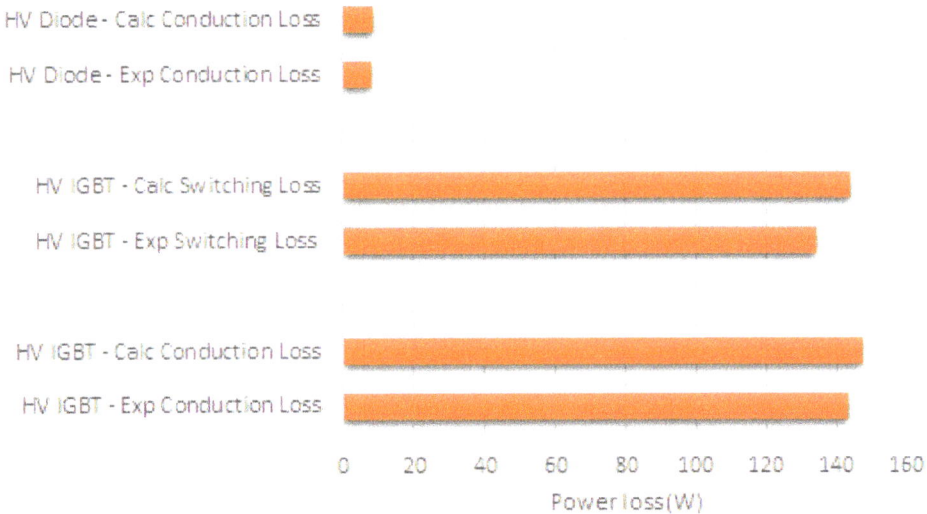

Figure 12. *Power loss comparison of HV-side devices—calculated (Calc) and experimental (Exp).*

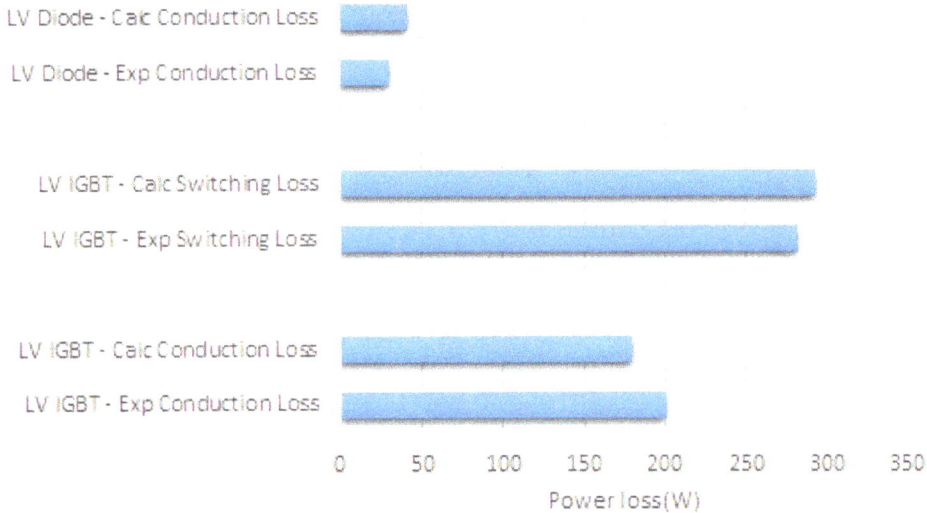

Figure 13. *Power loss comparison of LV-side devices—calculated (Calc) and experimental (Exp).*

Conclusion

Power loss analysis of IGBTs in a high-voltage high-power DAB DC-DC converter intended for use in an aerospace energy storage system was presented. The guidelines for selecting the appropriate IGBTs for a 20 kW, 540/125 V, 20 kHz DAB DC-DC converter prototype suitable for aerospace applications were given. The important parameters provided in the device datasheet for calculating the device power losses were discussed. Power loss analysis was performed for HV IGBTs corresponding to five leading manufacturers and LV IGBTs corresponding to four leading manufacturers based on the DAB converter prototype design. Apart from the guidelines given in this chapter, some IGBT manufacturers offer customer support for device loss estimation through a software package. Such software packages may be used for preliminary studies to estimate the device losses and the junction temperature for various operating conditions. In this work, experimental results were presented for 540 V, 80 A peak current operation on the HV-side IGBT, and 125 V, 300 A peak current on the LV-side IGBT. Using snubber capacitors on the HV side of the converter resulted in a nearly 45% reduction of IGBT switching losses.

Introducing snubber capacitors on the LV-side devices created parasitic ringing, with no significant reduction of switching power losses. However, snubbers on the LV side did reduce the device stresses by limiting dv/dt. The experimental results demonstrate that the use of snubber capacitors across IGBTs reduces switching losses and device stresses and thus improves converter performance.

Acknowledgements

The author thanks Rolls-Royce plc and the Engineering and Physical Sciences Research Council (EPSRC), UK, for the DHPA scholarship at the University of Manchester.

Author details

Thaiyal Naayagi Ramasamy

School of Electrical and Electronic Engineering, Newcastle University International Singapore, Singapore

*Address all correspondence to: naayagi.ramasamy@ncl.ac.uk

References

[1] DeDoncker RWRW, Divan DM, Kheraluwala MH. A three-phase soft- switched high-power-density dc/dc converter for high power applications. IEEE Transactions on Industry Applications. 1991;**27**(1):797-806. DOI: 10.1109/28.67533

[2] Kheraluwala MH, Gascoigne RW, Divan DM, Baumann ED. Performance characterization of a high-power dual active bridge dc-to dc converter. IEEE Transactions on Industry Applications. 1992;**28**(6):1294-1301. DOI: 10.1109/28.175280

[3] Segaran D, Holmes DG, McGrath BP. Comparative analysis of single and three-phase dual active bridge bi-directional dc-dc converters. In: Proc. Australian Universies Power Engineering Conference; 2008. pp. 1-6

[4] Tripathi AK, Mainali K, Patel DC, Kadavelugu A, Hazra S, Bhattacharya S, et al. Design considerations of a 15-kV SiC IGBT-based medium-voltage high-frequency isolated DC–DC converter. IEEE Transactions on Industry Applications. 2015;**51**(4):3284-3294. DOI: 10.1109/TIA.2015.2394294

[5] Kadavelugu A, Bhattacharya S, Ryu S-H, Brunt EV, Grider D, Agarwal A, Leslie S. Characterization of 15 kV SiC n-IGBT and its application considerations for high power converters. In: Proc. IEEE Energy Conversion Congress and Expo; 2013. pp. 2528-2535

[6] Kadavelugu A, Bhattacharya S. Design considerations and development of gate driver for 15 kV SiC IGBT. In: Proc. IEEE Applied Power Electronics Conference and Expo; 2014. pp. 1494-1501

[7] Hazra S, Madhusoodhanan S, Bhattacharya S, Moghaddam GK, Hatua K. Design considerations and performance evaluation of 1200 V, 100 a SiC MOSFET based converter for high power density application. In: Proc. IEEE Energy Conversion Congress and Expo; 2013. pp. 4278-4285

[8] Wen H, Xiao W, Bin S. Non active power loss minimization in a bidirectional isolated DC–DC converter for distributed power systems. IEEE Transactions on Industry Applications. 2014;**61**(12):6822-6831. DOI: 10.1109/ TIE.2014.2316229

[9] Oggier GG, Garcia GO, Oliva AR. Switching control strategy to minimize dual active bridge converter losses. IEEE Transactions on Power Electronics. 2009;**24**(7):1826-1838. DOI: 10.1109/ TPEL.2009.2020902

[10] Wang Y, de Haan SWH, Ferreira JA. Potential of improving PWM converter power density with advanced components. In: Proc. 13th European Conference on Power Electronics and Applications; 2009. pp. 1-10

[11] Gao Y, Huang AQ, Krishnaswami S, Richmond J, Agarwal AK. Comparison of Static and Switching Characteristics of 1200 V 4H-SiC BJT and 1200 V Si-IGBT. IEEE Transactions on Industry Applications. 2008;**44**(3):887-893. DOI: 10.1109/TIA.2008.921408

[12] Nomura T, Masuda M, Ikeda N, Yoshida S. Switching characteristics of GaN HFETs in a half bridge package for high temperature applications. IEEE Transactions on Power Electronics. 2008;**23**(2):692-697. DOI: 10.1109/ TPEL.2007.915671

[13] Semikron. Operation Principle of Power Semiconductors-Application Manual; 2008

[14] Semikron. Calculation of the Junction Temperature-Hints for Application-3.2.22008. p. 146

[15] Fuji Electric. Fuji IGBT Modules Application Manual; 2004

[16] Naayagi RT. Selection of power semiconductor devices for the DAB DC-DC converter for aerospace applications. In: Proc. IEEE 11th International Conference on Power Electronics and Drive Systems; 2015. pp. 499-502

[17] Advanced Power Technology. IGBT Tutorial Application Note APT0201. Rev. B. 2002

[18] Microsemi. Advanced IGBT Driver Application Manual-Application Note 1903; 2006

[19] Powerex. General Considerations for IGBT and Intelligent Power Modules- Application Note A10; 2008

[20] International Rectifier. Application Characterization of IGBTs-Application Note AN-990, rev.2; 2008

[21] IXYS. Choosing the Appropriate Component from Data Sheet Ratings and Characteristics-Technical information IXAN0056; 2008

[22] Ramasamy TN. Bidirectional DC-DC converter for aircraft electric energy storage systems [PhD thesis]. School of Electrical and Electronic Engineering, University of Manchester; 2010

[23] International Rectifier. IGBT or MOSFET: Choose Wisely-Application note; 2008

[24] Semikron. Power Modules: Special Features of Multi-Chip Structures- Application Note-1.4, AN1404; 2014

[25] Semikron. New Low-Inductive IGBT Module Constructions for High Currents and Voltages-Application Note-1.5.4; 2014

[26] Semikron. New Developments in MOSFET and IGBT Technology- Application Note-1.2.4;2014

[27] Naayagi RT, Forsyth AJ, Shuttleworth R. High-power bidirectional DC-DC converter for aerospace applications. IEEE Transactions on Power Electronics. 2012;**27**(11):4366-4379. DOI: 10.1109/ TPEL.2012.2184771

[28] Semikron. Ultrafast IGBT Module SK-M300GB125D Datasheet; 2009

[29] Semikron. Trench IGBT Module SKM600GB066D Datasheet; 2009

9

Electrical Field Distribution along HVDC GIL Spacer in SF$_6$/N$_2$ Gaseous Mixture

Boxue Du, Jin Li and Hucheng Liang

Abstract

Many researchers have proposed a variety of mathematical models to simulate the surface charge accumulation process of DC-GIS/GIL spacers. However, few of them took the gas collision ionization and charge trapping-detrapping process into consideration. This chapter combined the plasma hydrodynamics and charge transport equations and built a modified model. Some conclusions are shown as follows: for the basin-type spacer, the surface charge has the same polarity as the applied voltage on the lower surface but the opposite polarity on the upper surface. For the disc-type spacer, the surface charge has the same polarity as the applied voltage near the shell but the opposite polarity near the conductor under negative voltage. But under positive voltage, negative charge exists almost on the whole surface. The most serious distortion of the electric field occurs at the triple junction of epoxy spacer. Under load condition, there is an obvious temperature rise on the conductor due to joule heating, which has a great influence on the electric field distribution. The application of shielding electrodes has the function of field grading at the triple junction, which can be referred in the DC GIS/GIL design.

Keywords: DC-GIS/GIL, epoxy spacer, temperature, surface charge accumulation, electric field distribution, shielding electrode

Introduction

Over the last few decades, the high voltage direct-current (HVDC) transmission technology have been developed to carry out long distance and high capacity power transmission [1, 2]. Thanks to their high reliability and low footprint, Gas Insulated Switchgears and Gas Insulated Lines (GIS/GIL) have been widely used in AC power system but scarcely in DC system [3, 4]. One important reason is that surface charge is easy to accumulate on the spacer surface under DC voltage, which leads to a substantial electric field distortion and even flashover faults [5].

Therefore, the surface charge accumulation and electric field distribution along the GIS/ GIL spacers need to be understood.

In order to improve the reliability of GIS/GIL, researches aiming on the surface charge measurement and simulation have been conducted worldwide [6–8]. Some researchers built a scaled-down GIS/GIL and measured the surface charge distribution on the spacers [9–12]. Their measurements were conducted offline because the applied voltage on the conductor can easily cause damage to the electrostatic probe. The poor accuracy of offline measurement makes their measurements differ from each other. Besides, the accumulated surface charge at the triple junction can hardly be detected by the probe, which actually has a very significant effect on the electric field distribution.

Simulation is another way to study the surface charge distribution on spacers. Many researchers have proposed a variety of mathematical models to simulate the surface charge accumulation process [13–17]. However, few of them took the gas collision ionization and charge trapping-de-trapping process into consideration. This paper detailed the corona discharge and charge trapping-detrapping process by combining the plasma hydrodynamics and charge trapping-detrapping equations together. Under load conditions, GIS/GIL has an obvious temperature rise on the conductor. Considering there are many temperature-dependent factors in the surface charge accumulation process, the heat transfer equations are also coupled in this model to obtain the temperature distribution. Besides, the effects of shielding electrodes on the distorted electric field distribution are also studied.

Mathematical model

Geometric model

Figure 1 shows the geometric models of GIS with different spacer structures. There are two commonly used types of spacers as shown in **Figure 1a** which is shaped like a basin and **Figure 1b** which is shaped like a disc. In the simulation, the external diameter of the central conductor is set to 60 mm and the internal diameter of the shell is set to 330 mm. DC ±320 kV is applied on the central conductor and the shell is grounded.

Corona discharge in SF$_6$/N$_2$ mixture

In this paper, the SF$_6$/N$_2$ mixture is set as the insulating gas in GIS. The per- centage of SF$_6$ is set to 50% and the pressure is set to 0.5 MPa.

The drift-diffusion process of the electron density n_e and electron energy density n_ε can be described as [18],

$$\frac{\partial}{\partial t}(n_e) + \nabla \cdot \left[-n_e \left(\mu_e \cdot \vec{E} \right) - D_e \cdot \nabla n_e \right] = R_e \tag{1}$$

$$\frac{\partial}{\partial t}(n_\varepsilon) + \nabla \cdot \left[-n_\varepsilon \left(\mu_\varepsilon \cdot \vec{E} \right) - D_\varepsilon \cdot \nabla n_\varepsilon \right] + E \cdot \Gamma_e = R_\varepsilon \tag{2}$$

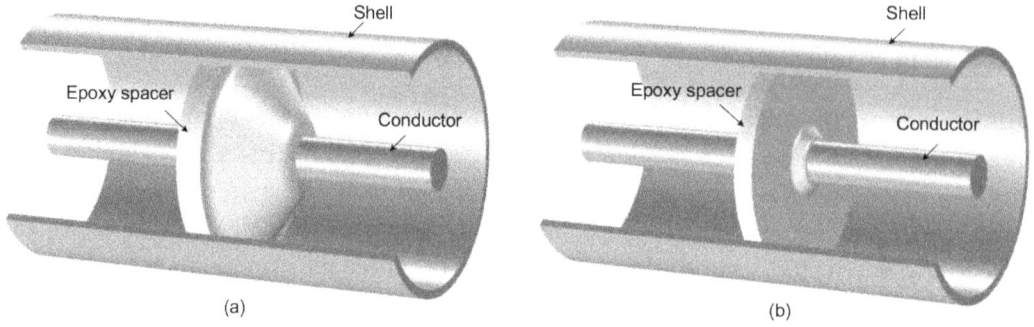

Figure 1. *The geometric models of GIS with different spacer structures: (a) basin-type and (b) disc-type.*

where the electron mobility μ_e, electron diffusivity D_e, energy mobility μ_ε and energy diffusivity D_ε are computed by solving the Two-term Boltzmann equation [19]. The electron source R_e and energy source R_ε can be obtained from Eqs. (3) and (4). This paper takes into consideration M electron-participating reactions (N two-body reactions and M-N three body reactions) and P energy-losing reactions,

$$R_e = \sum_{j=1}^{N} x_j k_j N_n n_e \Delta n_{ej} + \sum_{j=N+1}^{M} x_{j1} x_{j2} k_j N_n^2 n_e \Delta n_{ej} \tag{3}$$

$$R_\varepsilon = \sum_{j=1}^{P} x_j k_j N_n n_e \Delta \varepsilon_j \tag{4}$$

$$N_n = p/k_B T \tag{5}$$

where x_j, x_{j1} and x_{j2} are the mole fractions of the target species for reaction j; N_n is the total neutral density ($1/m^3$); p is the gas pressure (Pa); T is the temperature (K); Δn_{ej} is the electron increment of reaction j; $\Delta \varepsilon_j$ is the energy loss of reaction j (V) and k_j is the rate coefficient for reaction j (m^3/s), which is also obtained from the two-term Boltzmann equation [19].

For heavy species, the space–time dependent mass fraction of species k is controlled by the following equations,

$$\rho \frac{\partial}{\partial t}(w_k) - \nabla \cdot \vec{j_k} = R_k \tag{6}$$

$$\vec{j_k} = \rho w_k \vec{V_k} \tag{7}$$

$$\vec{V_k} = D_k \left(\frac{\nabla w_k}{w_k} + \frac{\nabla M_n}{M_n} \right) - z_k \mu_k \vec{E} \tag{8}$$

where ρ is the gas density (kg/m^3); w_k is the mass fraction of species k; j_k is the flux of species k; R_k is the rate expression for species k ($kg/(m^3 \cdot s)$); V_k is the multicomponent diffusion velocity of species k (m/s); D_k is the mixture averaged diffusion coefficient (m^2/s); M_n is the mean molar mass of the gas mixture (kg/ mol); z_k is the charge number of species k; μ_k is the mixture averaged mobility of species k ($m^2/(V\,s)$); E is the electric field strength (V/m). This paper takes into consideration M reactions (N two-body reactions and M-N three-body reactions) that change the mass fraction of species k,

$$R_k = \frac{M_k}{N_A}\left(\sum_{j=1}^{N} k_j x_{j1} x_{j2} N_n{}^2 \Delta n_{kj} + \sum_{j=N+1}^{M} k_j x_{j1} x_{j2} x_{j3} N_n{}^3 \Delta n_{kj}\right) \tag{9}$$

$$x_k = \frac{M_n w_k}{M_k} \tag{10}$$

where x_{j1}, x_{j2} and x_{j3} are the mole fractions of the species involved in reaction j; N_A is the Avogadro constant; Δn_{kj} is the species k increment of reaction j; M_k is the molar mass of species k (kg/mol). Eq. (9) shows the relationship between the mole fraction and mass fraction of species k.

The mixture averaged diffusion coefficient D_k is defined as Eq. (11) and the mixture averaged mobility μ_k can be calculated by Eq. (12) according to the Einstein's relation,

$$D_k = \frac{1 - w_k}{\sum_{j \neq k}^{Q} x_j / D_{k,j}} \tag{11}$$

$$\mu_k = \frac{e D_k}{k_B T} \tag{12}$$

where e is the unit charge (C); k_B is the Boltzmann's constant (J/K); T is the gas temperature (K) and $D_{k,j}$ is the binary diffusion coefficient between species k and j.

When electrons reach the conductor, the shell and the spacer surface, the boundary conditions for the electrons and electron energy flux can be defined as,

$$\vec{n} \cdot \vec{\Gamma}_e = \frac{1}{4} v_{e,th} n_e - \sum_p \gamma_p \left(\vec{\Gamma}_p \cdot \vec{n}\right) \tag{13}$$

$$\vec{n} \cdot \vec{\Gamma}_\varepsilon = \frac{5}{12} v_{e,th} n_\varepsilon - \sum_p \varepsilon_p \gamma_p \left(\vec{\Gamma}_p \cdot \vec{n}\right) \tag{14}$$

$$v_{e,th} = \left(\frac{8 k_B T_e}{\pi m_e}\right)^{1/2} \tag{15}$$

$$T_e = \frac{2 n_\varepsilon}{3 n_e} \tag{16}$$

The second term on the right-hand side of Eqs. (13) and (14) are the gain of electrons and electron energy due to secondary emission effects; γ_p is the secondary emission coefficient, which is set to 0.1 for the cathode and 0 for the anode; ε_p is the mean energy of the secondary electrons, which is set to 5 eV; $v_{e,th}$ is the electron thermal velocity (m/s); T_e is the electron temperature (eV).

For ions, the boundary condition can be described as,

$$\vec{n} \cdot \vec{j}_k = \frac{1}{4} v_{k,th} \rho w_k + \alpha \rho w_k z_k \mu_k \left(\vec{E} \cdot \vec{n}\right) \tag{17}$$

$$v_{k,th} = \left(\frac{8 k_B T}{\pi m_k}\right)^{1/2} \tag{18}$$

where $v_{k,th}$ is the thermal velocity of species k (m/s); T can be considered to be the gas temperature in GIS (K); $a \bullet z_k / |z_k| = 0$ if the electric field is directed away from the boundary and $a \bullet z_k / |z_k| = 1$ if the electric field is directed toward the boundary.

The discharge processes of SF_6/N_2 mixture are pretty complex and some of them are still unclear. In this simulation, only those primary reactions are taken into consideration. **Table 1** lists some typical physicochemical reactions considered in this paper after some reduction. The cross sections and energy losses of those collision reactions are extracted from papers [20–23].

Charge transport in epoxy spacer

When ions and electrons reach the spacer surface, charges are considered to be injected into the skin layer of epoxy owing to the surface reactions ($N+_2 \rightarrow N_2$, $SF_6^+ \rightarrow SF_6$, $SF_6^- \rightarrow SF_6$) [14]. The charge transport process in epoxy spacer volume is mainly controlled by the following equations (19)-(21) [24–26]. Note that the epoxy spacer in this numerical model is considered to be clean enough, the surface current due to surface transmission can be ignored since there is no special surface treatment to the spacer [27]. Therefore, the surface transmission process is not taken into consideration in this paper.

Table 1. *Table of some physicochemical reactions in the corona discharge.*

No.	Formula	Type	$\Delta \varepsilon$ (eV)	Δn_e
1	$N_2 + e \rightarrow 2e + N_2^+$	Ionization	−16	1
2	$SF_6 + e \rightarrow 2e + SF_6^+$	Ionization	−15.8	1
3	$N_2 + e \rightarrow e + N_2$	Excitation	−8	0
4	$SF_6 + e \rightarrow e + SF_6$	Excitation	−10	0
5	$N_2 + e \rightarrow e + N_2$	Elastic	0	0
6	$SF_6 + e \rightarrow e + SF_6$	Elastic	0	0
7	$SF_6 + e \rightarrow SF_6^-$	Attachment	—	−1
8	$SF_6^+ + e \rightarrow SF_6$	Reaction	—	−1
9	$SF_6^+ + SF_6^- \rightarrow 2SF_6$	Reaction	—	—
10	$e + N_2^+ + N_2 \rightarrow 2N_2$	Reaction	—	−1
11	$2e + N_2^+ \rightarrow N_2 + e$	Reaction	—	−1

$$\frac{\partial n_{mb}}{\partial t} + \nabla \cdot \vec{J}_c^e = n_{tr}P_{de} - n_{mb}P_{tr} - R_1 n_{mb}h_{tr} - \frac{1}{\tau}\Delta n \tag{19}$$

$$\frac{\partial n_{tr}}{\partial t} = n_{mb}P_{tr} - R_2 n_{tr}h_{mb} \tag{20}$$

$$\vec{J}_c^e = \mu_{e,2}n_{mb} \cdot \nabla V - D_{e,2} \cdot \nabla n_{mb} \tag{21}$$

$$\frac{\partial h_{mb}}{\partial t} + \nabla \cdot \vec{J}_c^h = h_{tr}P_{de} - h_{mb}P_{tr} - R_2 n_{tr}h_{mb} - \frac{1}{\tau}\Delta h \tag{22}$$

$$\frac{\partial h_{tr}}{\partial t} = h_{mb}P_{tr} - R_1 n_{mb}h_{tr} \tag{23}$$

$$\vec{J}_c^h = -\mu_h h_{mb} \cdot \nabla V - D_h \cdot \nabla h_{mb} \tag{24}$$

where n_{mb} is the mobile electron density (1/m^3); n_{tr} is the trapped electron density (1/m^3); h_{mb} is the mobile hole density (1/m^3); h_{tr} is the trapped hole density (1/m^3); $\vec{J}_c^e$ and $\vec{J}_c^h$ are the flux of mobile electrons and holes (1/(m^2•s)); $\mu_{e,2}$ and μ_h are the mobility of electrons and holes in epoxy spacer (m^2/(V•s)); $D_{e,2}$ and D_h are the diffusion coefficient of electrons and holes (m^2/s); V is the potential (V); P_{tr} and P_{de} are the trapping and de-trapping coefficient of electrons and holes (1/s); Δn and Δh are the non-equilibrium carrier density in epoxy spacer (1/m^3); R_1 and R_2 are the recombination coefficients (m^3/s); τ is the lifetime of non-equilibrium carriers (s).

In this paper, the volume conductivity, carrier mobility and diffusion coefficient of spacer are assumed as following:

$$\sigma = 40000 \cdot e^{-12860/T} \tag{25}$$

$$\mu_{e,2} = \mu_h = (1.5 \times 10^{-4}) \cdot e^{-6470/T} \tag{26}$$

$$D_{e,2} = D_h = \mu_{e,2} \cdot \frac{k_B T}{e} \tag{27}$$

where T is the temperature of spacer (K). Supposing the mobile electron density n_0 and hole density h_0 have the same value, the intrinsic carrier density can be described as,

$$n_0 = h_0 = \frac{\sigma}{2e\mu_{e,2}} \tag{28}$$

Inspired from the theories on the non-equilibrium carriers in the semiconductor physics, this paper considers the product of n_0 and h_0 as a constant in the epoxy spacer under thermal equilibrium condition. Supposing the electron and hole density n_{mb} and h_{mb} are in an unbalanced state, Δn and Δh can be obtained by solving the Eq. (29). Positive Δn and Δh represent the electron–hole recombination process, negative Δn and Δh represent the generation of electron–hole pairs.

$$n_0 h_0 = (n_{mb} - \Delta n)(h_{mb} - \Delta h) \tag{29}$$

$$\Delta n = \Delta h = \frac{1}{2} \left[(n_{mb} + h_{mb}) - \sqrt{(n_{mb} - h_{mb})^2 + 4n_0 h_0} \right] \tag{30}$$

At the gas–solid interface, the boundary condition can be described as,

$$-\vec{n} \cdot \vec{\Gamma}_{n_{mb}} = \vec{n} \cdot \vec{j}_e \tag{31}$$

$$-\vec{n} \cdot \vec{\Gamma}_{h_{mb}} = \vec{n} \cdot \vec{j}_i \tag{32}$$

where $\vec{J}_e$ and $\vec{J}_i$ is the electron and ion flux through the gas–solid interface in the corona discharge model (1/m^2•s).

At the metal-insulation interface, the boundary condition can be described as,

$$- \vec{n} \cdot \vec{j_e} = \alpha \cdot J_{sch-e} - (1 - \alpha) \cdot \vec{n} \cdot \vec{E} \cdot \mu_{e,2} \cdot n_{mb} \tag{33}$$

$$- \vec{n} \cdot \vec{j_h} = (1 - \alpha) \cdot J_{sch-h} - \alpha \cdot \vec{n} \cdot \vec{E} \cdot \mu_h \cdot h_{mb} \tag{34}$$

$$J_{sch-h} = J_{sch-h} = \frac{AT^2}{e} \cdot \exp\left(\frac{-\phi + \sqrt{E \cdot e^3 / 4\pi\varepsilon_0\varepsilon_r}}{k_B T} \right) \tag{35}$$

where J_{sch-e} and J_{sch-h} are the Schottky injection electron and hole flux (1/(m²•s)); A is the Richardson constant (C/m²·s·K²), ϕ is the contact potential barrier between the spacer and metal (eV); $\alpha = 0$ if the electric field is directed away from the boundary and $\alpha = 1$ if the electric field is directed toward the boundary. Table 2 lists some parameters of epoxy spacer used in the charge transport model.

Table 2. *Table of some parameters in the charge transport model.*

Parameters	Value
Trapping and de-trapping coefficients, 1/s	
P_{tr}	7.0×10^{-3}
P_{de}	7.7×10^{-5}
Recombination coefficients, m³/s	
R_1	8.0×10^{-19}
R_2	8.0×10^{-19}
Potential barrier, eV	
ϕ	1.25

Poisson equation

The corona discharge model comprises p species (electrons, positive ions and negative ions) and the charged number of specie k is z_k. Thus, in the SF_6/N_2 gas, Poisson equation can be described as,

$$-\nabla^2 V = e \left(\sum_1^p n_k z_k \right) / \varepsilon_0 \tag{36}$$

where e is the elementary charge (C); n_k is the density of species k (1/m³); ε_0 is the vacuum dielectric constant (F/m).

In the epoxy spacer, Poisson equation is described as,

$$-\nabla^2 V = e(h_{mb} + h_{tr} - n_{mb} - n_{tr})/(\varepsilon_0\varepsilon_r) \tag{37}$$

where ε_r is the relative dielectric constant of epoxy spacer, which is set to 4.

Heat transfer in GIS

In GIS, the heat transfer is controlled by the following equation [28],

$$\rho C_p \frac{\partial T}{\partial t} + \rho C_p \vec{u} \cdot \nabla T = \nabla \cdot (k \nabla T) + Q_\rho, \tag{38}$$

$$\rho \frac{\partial \vec{u}}{\partial t} + \rho \left(\vec{u} \cdot \nabla \right) \vec{u} = \nabla \cdot \left[-p \vec{I} + \mu \left(\nabla \vec{u} + \left(\nabla \vec{u} \right)^T \right) - \frac{2}{3} \mu \left(\nabla \cdot \vec{u} \right) \vec{I} \right] + \vec{G} \tag{39}$$

$$\frac{\partial \rho}{\partial t} + \nabla \cdot \left(\rho \vec{u} \right) = 0 \tag{40}$$

where ρ is the density of SF$_6$/N$_2$ gas, epoxy spacer or shell (kg/m³); C_p is the heat capacity (J/(kg·K)); k is the thermal conductivity (W/(m·K)); T is the thermodynamic temperature (K); u is the natural convection velocity field of SF$_6$/N$_2$ gas (m/s); μ is the coefficient of kinetic viscosity, which is set to 1.7 x 10^{-5} Pa s; G is the gravity vector (N/m³); Q_ρ is the heat source in the spacer (W/m³), which is defined as,

$$Q_\rho = eE^2 \mu (n_{mb} + h_{mb}) \tag{41}$$

The density of SF$_6$/N$_2$ gas in GIS is defined as,

$$\rho = \frac{M_n p}{RT} \tag{42}$$

where M_n is the mean molar mass (kg/mol).

The heat radiation of conductor, spacer and shell is described as,

$$- \vec{n} \cdot (-k \nabla T) = \varepsilon \sigma \left(T_{amb}^4 - T^4 \right) \tag{43}$$

where σ is the Avogadro Boltzmann constant; T_{amb} is the environment temperature; ε is the radiation coefficient, which is set to 0.8 for the spacer, 0.05 for the conductor and 0.15 for the shell.

In addition to the heat radiation at the shell-air interface, the natural convection of air can be described as,

Table 3. *Table of some parameters in the heat transfer model.*

Parameters	Spacer	Conductor/shell	SF$_6$/N$_2$
Thermal conductivity k, W/(m K)	0.71	165	0.019
Heat capacity C_p, J/(kg K)	700	900	817
Density ρ, kg/m³	2200	2700	—

$$- \vec{n} \cdot (-k \nabla T) = h(T_0 - T) \tag{44}$$

where T_0 is the environment temperature (K) and h is the heat transfer coefficient, which is set to 10 W/(m²·K). Table 3 lists some parameters of GIS used in the heat transfer model.

Note that many simulation parameters such as $\mu_{e,2}$, μ_h, $D_{e,2}$, D_h, P_{de}, n_o, $v_{k,th}$, J_{sch-e}, J_{sch-h}, ρ and so on are all temperature-dependent. As a result, the temperature distribution has a great influence on the surface charge accumulation process. In this paper, the environment temperature is set to 293 K and the temperature of central conductor is respectively set to 330, 350 and 370 K.

Results and discussion

This paper built a temporal numerical model to simulate the surface charge accumulation process. But what should be cared about is the steady-state surface charge and electric field distributions. After many calculations, it can be drawn that the temporal change in surface charge density slows down with time under both negative and positive voltages. However, it takes a much longer time for the surface charge density to reach the steady state under negative voltage than that under positive voltage. For convenience, the simulation time is set to be 1 hour. After 1 hour, the increment of surface charge density has been slow enough. The difference between the surface charge distribution and the steady state is less than 10%. As a result, the surface charge density after 1 hour can be approximately considered as the steady-state distribution. It is necessary to note that it takes more than 20 hours for the measured surface charge distribution to reach the steady state [11, 12], which is much longer compared with the simulation. This phenomenon is mainly caused by the non-uniformity of material properties in the surface charge measurement.

Effects of surface charge

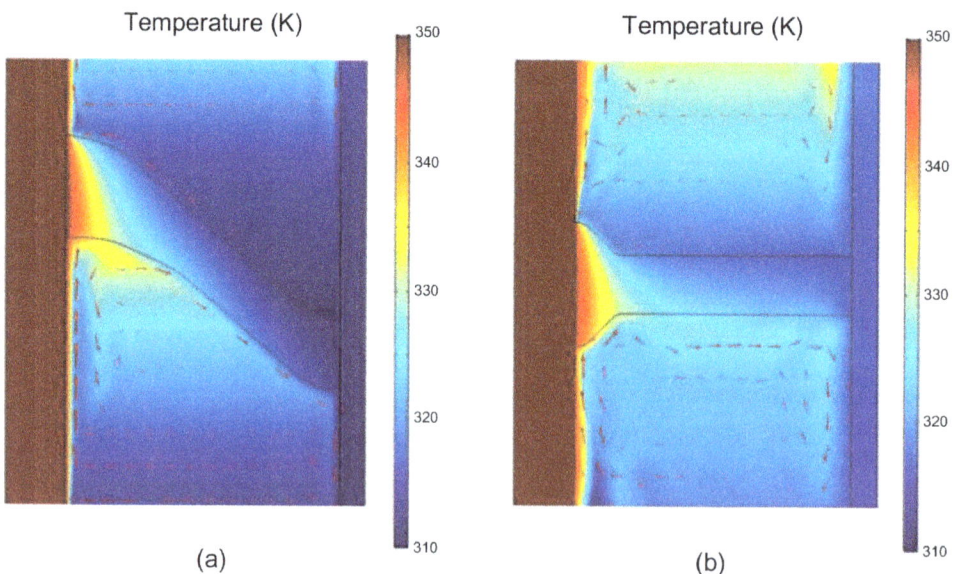

Figure 2. *The simulated temperature distribution and gas convection in DC-GIS.*

When GIS is under load condition, the central conductor has an obvious temperature rise due to joule heating. In this paper, the room temperature is set to 293 K. **Figure 2** shows the simulated temperature distribution and gas convection in DC-GIS when the conductor temperature is set to 350 K. The temperature goes down from the conductor to the shell and the shell temperature is about 17 K higher than the room temperature. Because of the gas convection, the spacer temperature is higher on the lower surface than that on the upper surface. There is a large temperature range along the epoxy spacers, thus it is important to study the temperature effects on the surface charge and electric field distribution.

Figure 3 shows the simulated surface charge distribution on the basin-type spacer, which is similar to the measured results in paper [9, 11]. The conductor temperature is set to 350 K. The color variation from red to blue represents the conversion of surface charge polarity from positive to negative. In this paper, the convexity and concave of spacer are defined as the upper and lower surface. Generally speaking, the surface charge has the same polarity as the applied voltage on the lower surface but the opposite polarity on the upper surface. However, a density drop or even a polarity reversal of surface charge shows up near the central conductor.

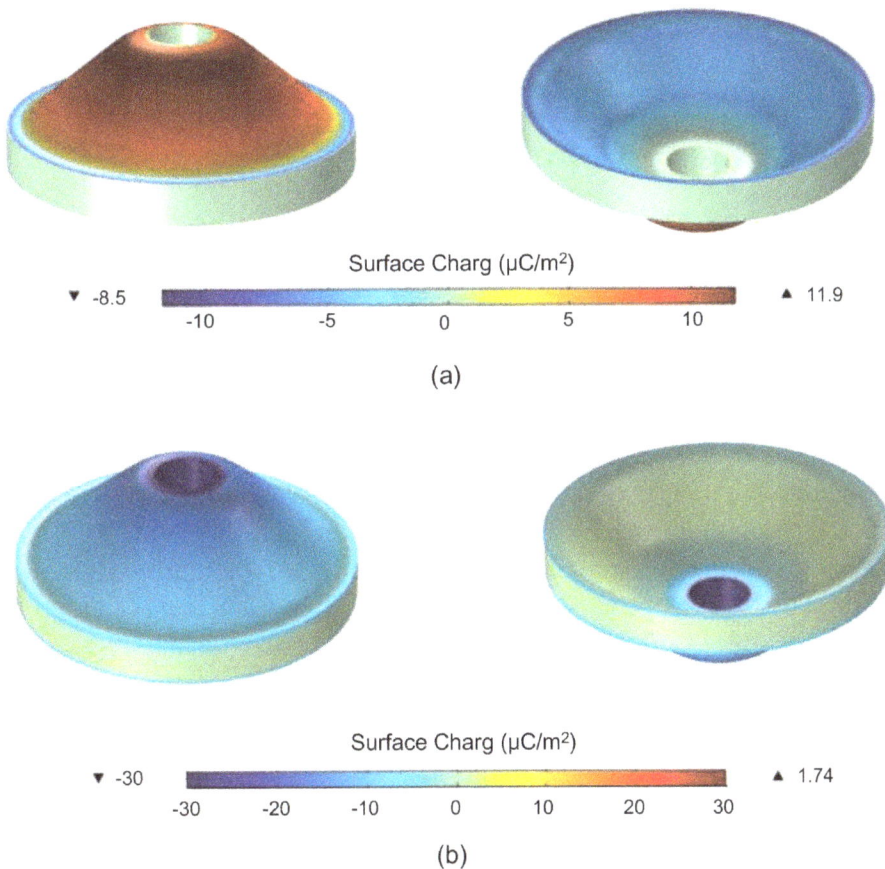

Figure 3. *The simulated surface charge distribution on the basin-type spacer at 350 K: (a) negative voltage and (b) positive voltage.*

Figure 4 shows the mechanism of the surface charge accumulation process, which can be used to explain the surface charge distribution in **Figure 3**. Usually, three processes are considered to influence the surface charge accumulation on GIS spacers. (1) Gas transmission: Ions and electrons produced by corona discharge in the insulating gas transport to the spacer surface by electric force; (2) Volume transmission: Carriers in the spacer volume transport to the spacer interface by electric force; (3) Surface transmission: Carriers transport along the spacer surface by electric force. In this paper, the spacer is considered to have a pretty clean surface, the surface transmission is thus not as influential as the gas and volume transmission.

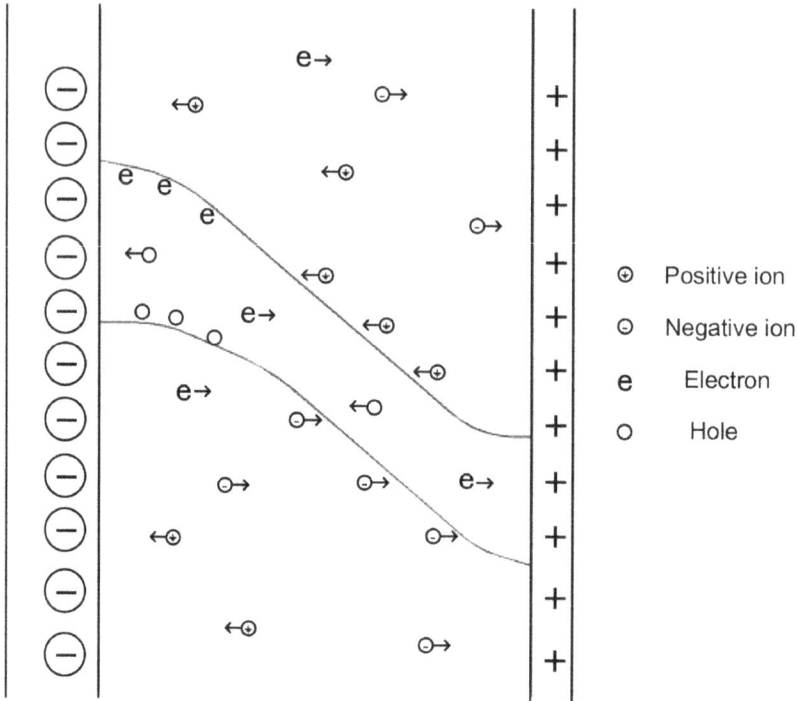

Figure 4. *Mechanism of the surface charge accumulation process.*

As shown in **Figure 4** under negative voltage, electrons and negative ions in the SF_6/N_2 mixture transport from the conductor to the shell and accumulate on the lower surface of spacer. Accordingly, the positive ions transport in the opposite direction and accumulate on the upper surface. In this case, the gas transmission takes the main part in the surface charge accumulation process. As a result, the surface charge has the same polarity as the applied voltage on the lower surface but the opposite polarity on the upper surface. When the conductor temperature is set to 350 K, there is a large temperature range along the spacer. According to the Eq. (25), the volume conductivity of spacer has been greatly increased near the central conductor. In this case, the volume transmission takes a more important role in the surface charge accumulation process. Electrons transport away from the conductor in the spacer volume and accumulate at the upper interface. Holes transport in the opposite direction and accumulate at the lower interface. This could explain why the density drop or the polarity reversal of surface charge shows up near the central conductor.

Figure 5 shows the electric field distribution along the basin-type spacer under both negative and positive voltages. The temperature of conductor is set to 350 K. The coordinate r from 30 to 167 mm represents the spacer surface from the central conductor to the shell. In this paper, the electric field distribution without corona discharge is defined as the normal field. And the field distribution with corona discharge is defined as the distorted field. It can be seen that the normal field near the conductor is much weaker than that in the surrounding zone. This is induced by the greatly enhanced volume conductivity of spacer due to temperature rise on the conductor. When the corona discharge is taken into consideration, ions and electrons in the SF$_6$/N$_2$ mixture will transport under electric force and accumulate on the spacer surface. Surface charge due to gas transmission can evidently distort the electric field. **Figure 5a** presents the distorted field distribution under negative voltage. The most serious distortion of electric field shows up in the zone close to the shell. From **Figure 3a** it can be learned that there is an annular zone with negative charge close to the shell. When negative voltage is applied on the conductor, the inner wall of the shell carries positive charges. Thus the existence of the negative charge zone greatly strengthens the electric field. **Figure 5b** presents the distorted field distribution under positive voltage. The most serious distortion of electric field shows up in the zone close to the conductor. Similarly, the negative charge zone surrounding the conductor should be responsible.

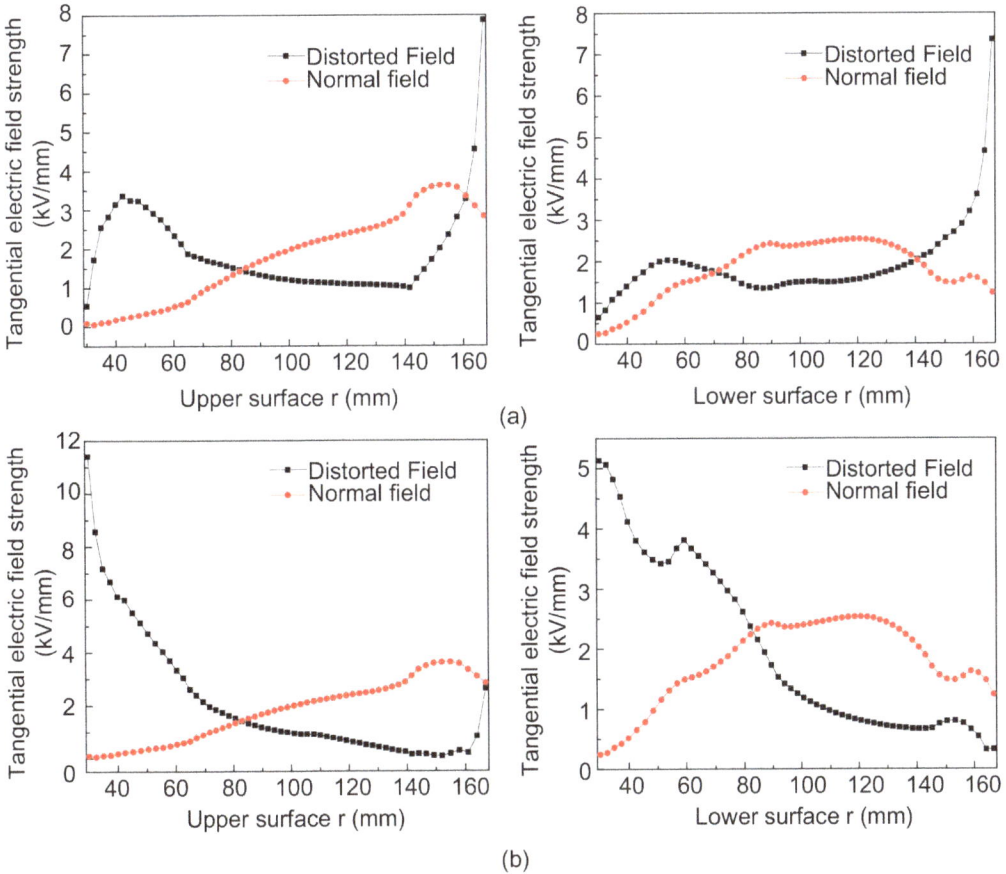

Figure 5. *The electric field distribution along the basin-type spacer at 350 K: (a) negative voltage and (b) positive voltage.*

Figure 6 shows the surface charge and the corresponding electric field distribution along the disc-type spacer under both negative and positive voltages. The temperature of conductor is set to 350 K. As shown in **Figure 6a** under negative voltage, the surface charge has the same polarity as the applied voltage near the shell but the opposite polarity near the conductor. But under positive voltage, negative surface charge exists almost on the whole surface. **Figure 6b** presents the normal and distorted field distribution under both negative and positive voltages. Without considering the corona discharge, the normal field near the conductor is weaker than that in the surrounding zone, which is similar as the field distribution along the basin-type spacer. The serious field distortion under negative voltage is induced by the negative charge zone close to the shell. And the serious field distortion under positive voltage is induced by the negative charge zone surrounding the conductor.

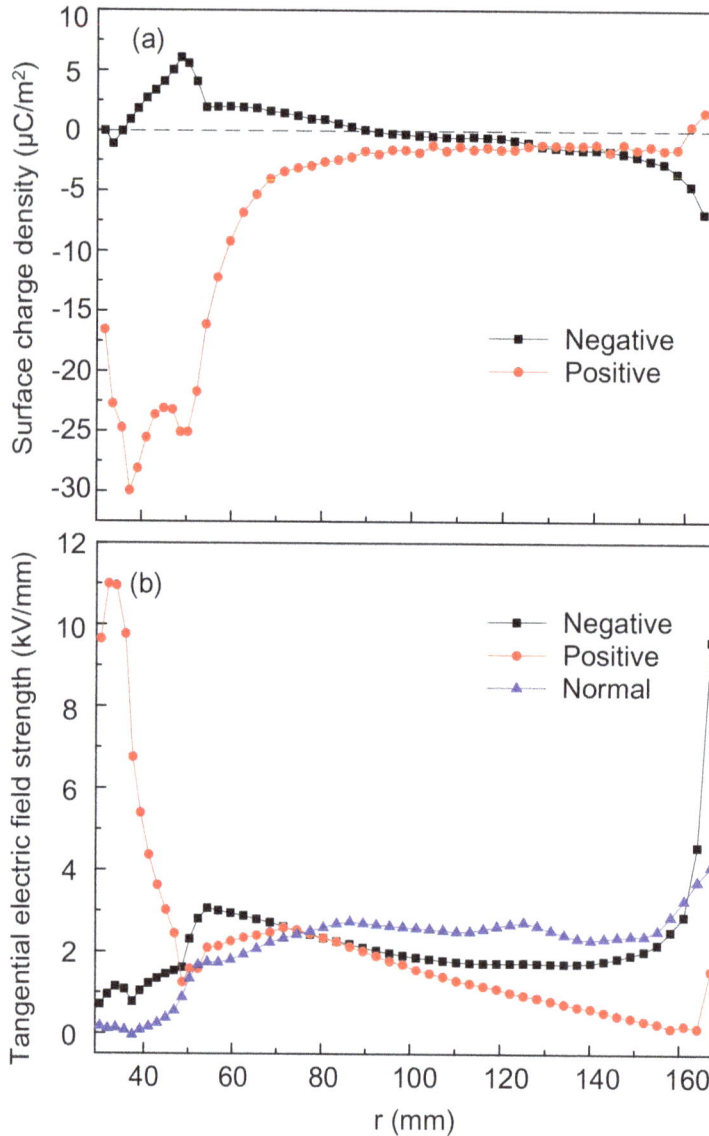

Figure 6. *Surface charge and the corresponding electric field distribution along the disc-type spacer at 350 K: (a) surface charge and (b) electric field.*

Effects of temperature

Under different load conditions, the conductor has different temperatures, leading to different temperature distributions in GIS. In this paper, many factors taken into consideration are temperature-dependent, such as carrier transport coefficient, carrier concentration, ion transport coefficient, Schottky injection current, gas density, thermal conductivity and so on. Therefore, temperature has a great influence on the surface charge accumulation process and the electric field distribution.

Figure 7 shows the surface charge distribution along the basin-type spacer at different temperatures. Under negative voltage as shown in **Figure 7a**, less positive charges accumulate on the upper surface as the temperature increases. On the lower surface, the density of negative surface charge decreases with temperature and more positive charges are observed near the conductor. As the temperature increases, the conductivity of epoxy spacer grows up sharply. The volume transmission takes a more important role in the surface charge accumulation process.

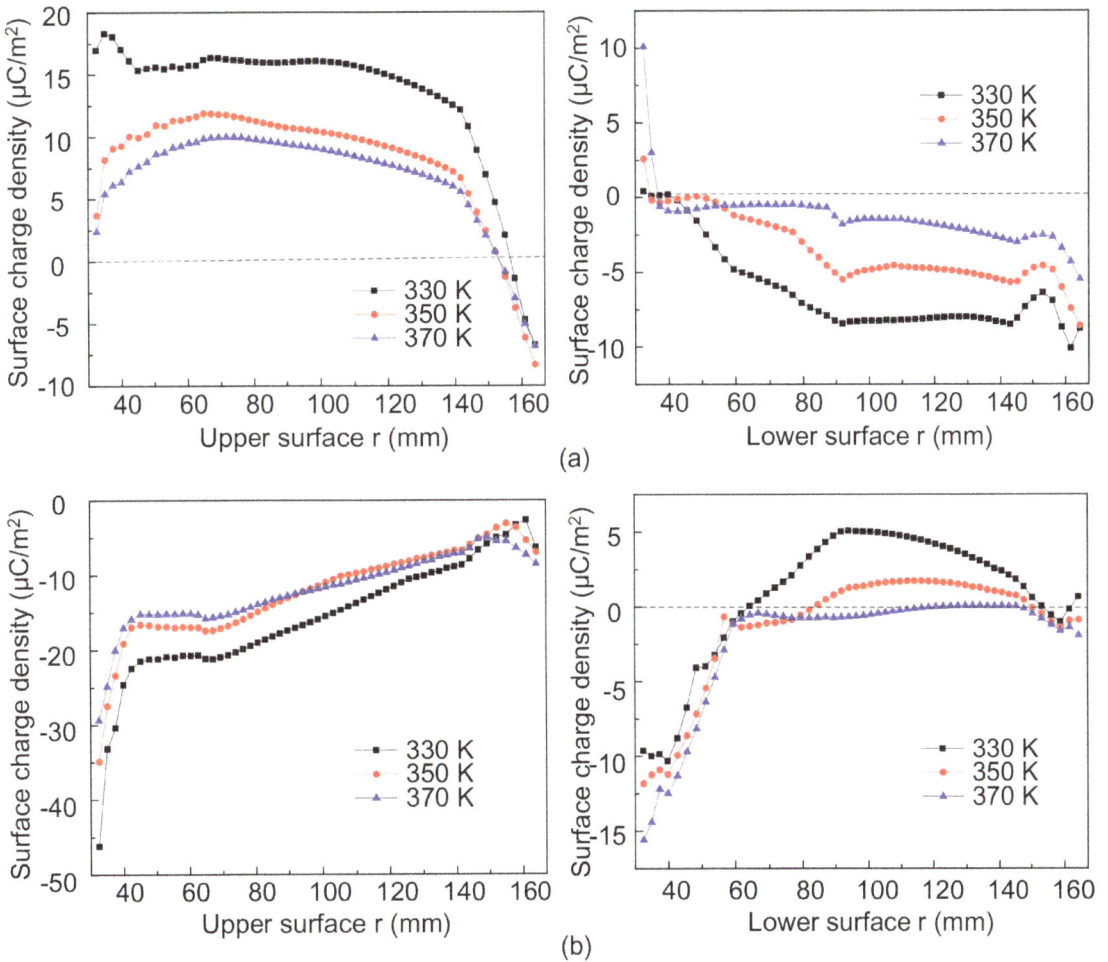

Figure 7. *The surface charge distribution along the basin-type spacer at different temperatures: (a) negative voltage and (b) positive voltage.*

Similarly, under positive voltage as shown in **Figure 7b**, the volume transmission causes more positive charges to transport to the upper surface. Thus the negative surface charge due to gas transmission decreases with temperature. On the lower surface, more negative charges and less positive charges are observed due to the enhanced volume transmissions as the temperature increases.

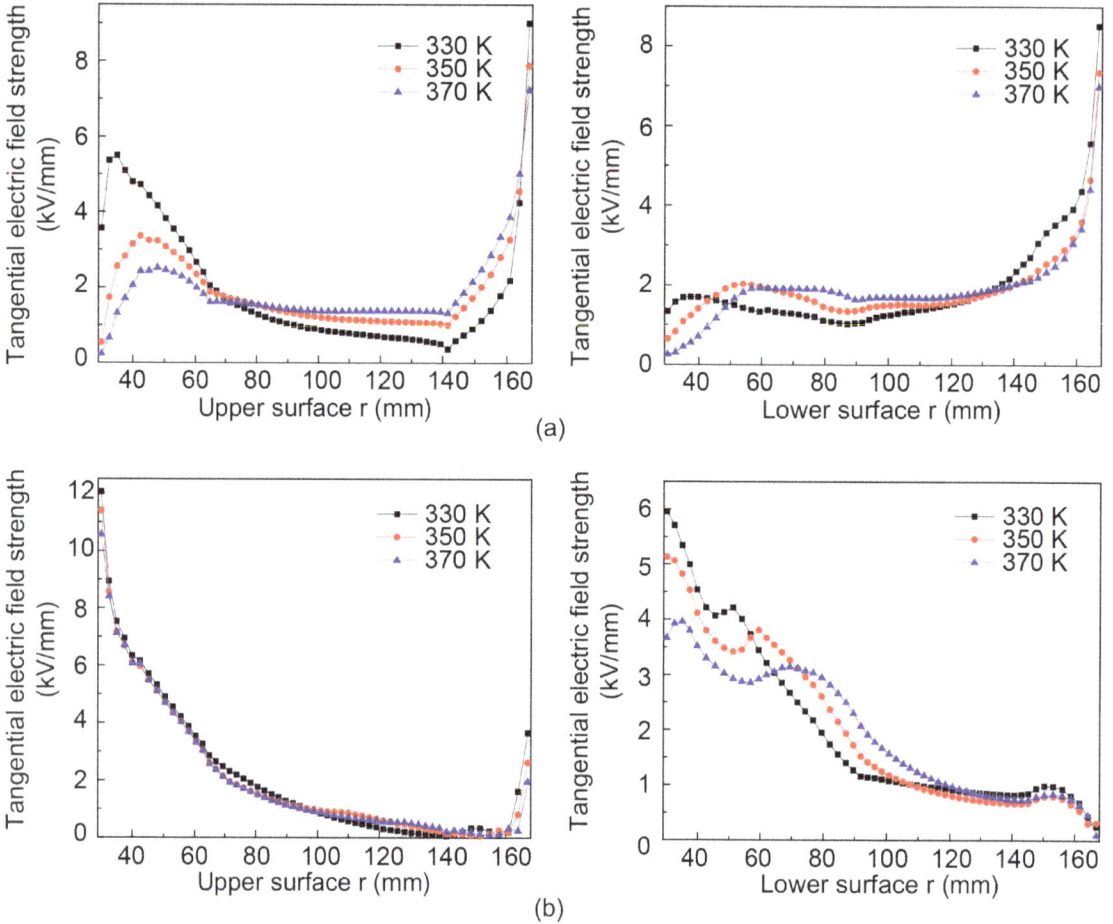

Figure 8. *The distorted electric field distribution along the basin-type spacer at different temperatures: (a) negative voltage and (b) positive voltage.*

Figure 8 shows the distorted electric field distribution along the basin-type spacer at different conductor temperatures. Under negative voltage as shown in **Figure 8a**, the field strength near the conductor is obviously decreased as the temperature increases. As the temperature goes up, the spacer has a huge temperature rise near the conductor, leading to the sharp increase in the volume conductivity. As a result, the grown volume conductivity greatly weakens the electric field near the conductor. Under positive voltage as shown in **Figure 8b**, the electric field near the conductor is also weakened with the increasing temperature, especially on the lower spacer surface. It seems that the heavy load on GIS is helpful to uniform the electric field distribution under positive voltage.

Figure 9 shows the distorted electric field distribution along the disc-type spacer at different conductor temperatures. Similarly, the electric field is weakened near the conductor with the increasing temperature under both negative and positive voltages. This is induced by the temperature-dependent volume conductivity of epoxy spacer. When positive voltage is applied on the disc-type spacer, the huge temperature rise on the conductor is also considered to have the effect of field grading.

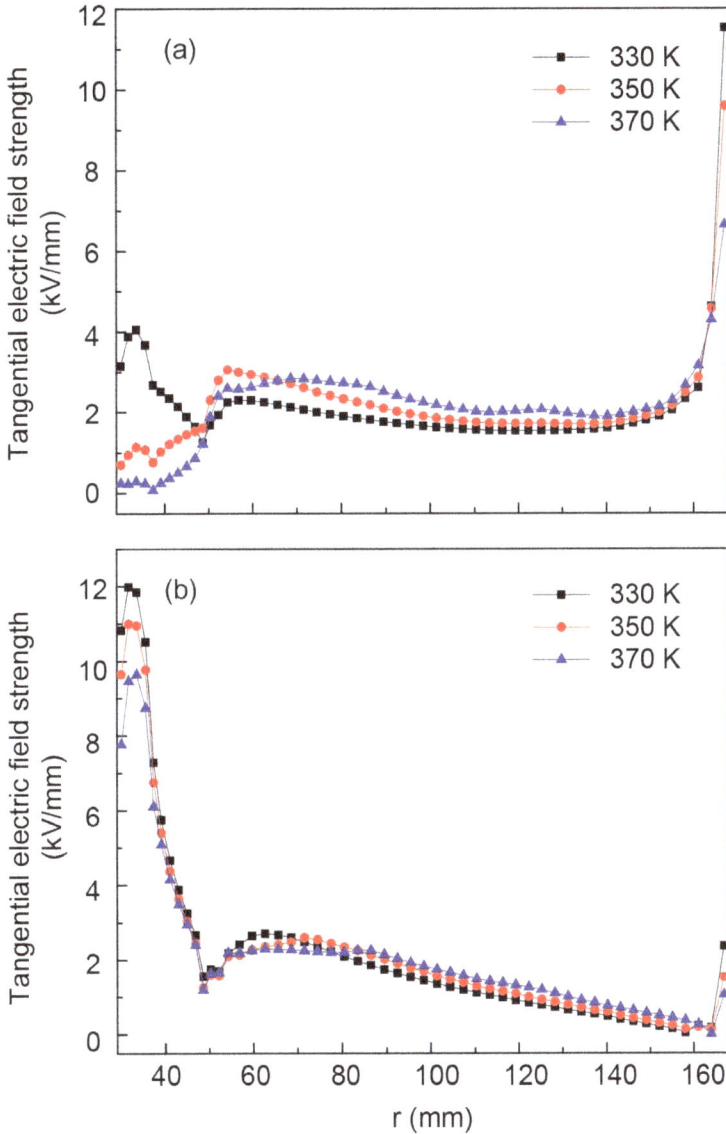

Figure 9. *The distorted electric field distribution along the disc-type spacer at different temperatures. (a) negative voltage, (b) positive voltage.*

Effects of shielding electrode

From the results in Section 3.1 it can be inferred that the most serious distortion of electric field occurs at the triple junction in GIS. On various occasions, the triple junction is always the cause of field distortion and partial discharge. In order to solve this problem, both the conductor and shell are designed with shielding structures as shown in **Figure 10**. Note that the grounded shielding electrode is designed to be concave on the shell. As a result, the distance along the spacer surface from the conductor to the shell becomes a little longer.

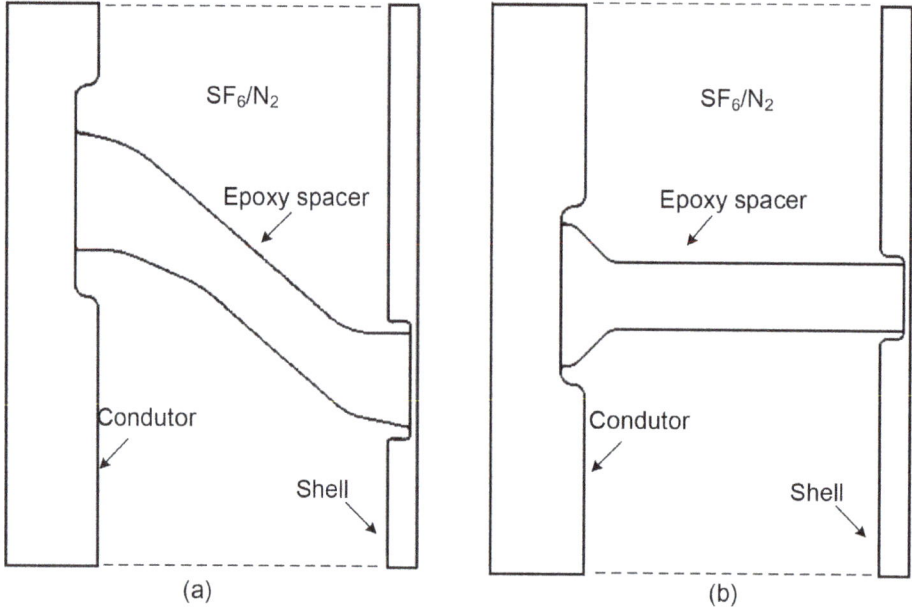

Figure 10. *GIS with shielding structures on the conductor and shell. (a) basin-type spacer, (b) disc-type spacer.*

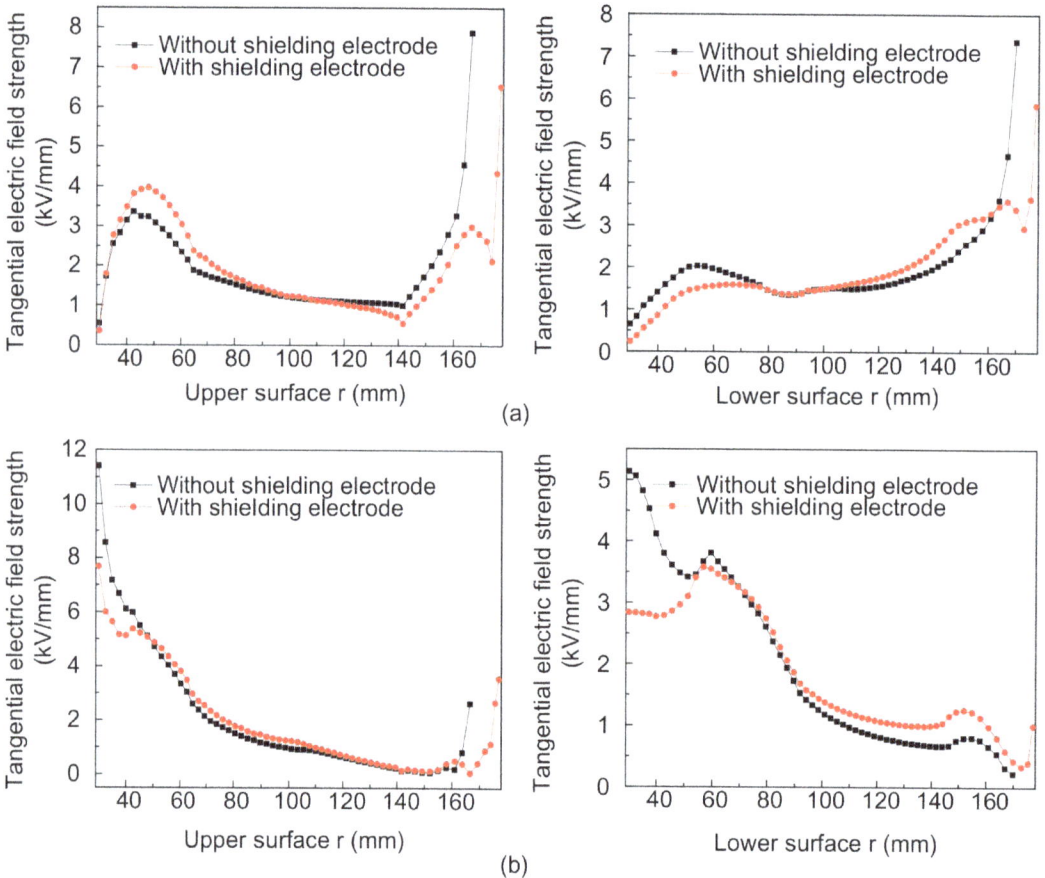

Figure 11. *The electric field distribution along the basin-type spacer at 350 K with and without shielding electrodes: (a) negative voltage and (b) positive voltage.*

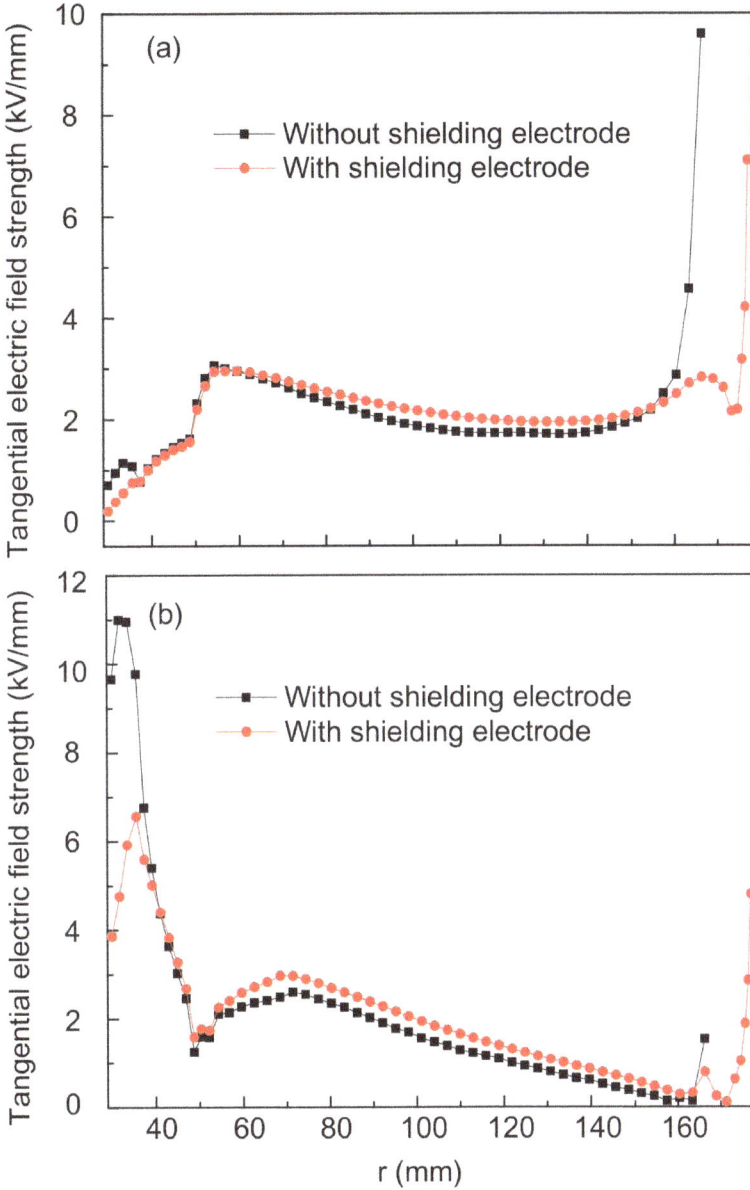

Figure 12. *The electric field distribution along the disc-type spacer at 350 K with and without shielding electrodes: (a) negative voltage and (b) positive voltage.*

Figure 11 presents the comparison of electric field between the GIS with and without shielding electrodes. It is easy to find that the electric field distribution is greatly improved by the shielding electrodes. As mentioned above, the most serious distortion of electric field is induced by the negative surface charges at the triple junction. The existence of shielding electrodes prevents the charges from reaching the spacer surface at the triple junction. Besides, the uniformed electric field at the triple junction is considered to suppress the partial discharge, less surface charges are produced. **Figure 12** presents the electric field distribution along the disc-type spacer with and without shielding electrodes. Also, the electric field distribution along the disc-type spacer is uniformed by the shielding electrodes.

Conclusions

This paper built a numerical model to simulate the surface charge accumulation process and the electric field distribution, the main conclusions can be drawn as follows:

1. Under load condition, there is a huge temperature rise on the GIS conductor due to joule heating. The wide variation of temperature along the epoxy spacers has a great influence on the surface charge accumulation process.

2. For the basin-type spacer, the surface charge has the same polarity as the applied voltage on the lower surface but the opposite polarity on the upper surface due to gas transmission. Because of the increasing spacer conductivity with temperature, the volume transmission takes a more important role in the surface charge accumulation process near the conductor. A density drop or even a polarity reversal of surface charge is observed near the conductor. For the disc-type spacer, the surface charge has the same polarity as the applied voltage near the shell but the opposite polarity near the conductor under negative voltage. But under positive voltage, negative charge exists almost on the whole surface.

3. The most serious distortion of the electric field distribution along the epoxy spacers occurs at the triple junction. As the conductor temperature increases, the field strength near the conductor is weakened owing to the temperature- dependent volume conductivity.

4. The application of shielding electrodes has the function of field grading at the triple junction, which can be referred in the DC-GIS/GIL design.

Author details

Boxue Du, Jin Li* and Hucheng Liang

Key Laboratory of Smart Grid of Education Ministry, School of Electrical and Information Engineering, Tianjin University, Tianjin, China

*Address all correspondence to: lijin@tju.edu.cn

References

[1] Delucchi MA, Jacobson MZ. Providing all global energy with wind, water, and solar power, Part II: Reliability, system and transmission costs, and policies. Energy Policy. 2011; 39(3):1170-1190. DOI: 10.1016/j. enpol.2010.11.045

[2] Liu Y, Li C, Mu Q, et al. Side-by-side connection of LCC-HVDC links to form a DC grid. In: European Conference on Power Electronics and Applications. IEEE. 2015. pp. 1-10. DOI: 10.1109/ EPE.2015.7311787

[3] Du BX, Li A. Effects of DC and pulse voltage combination on surface charge dynamic behaviors of epoxy resin. IEEE Transactions on Dielectrics & Electrical Insulation. 2017;24(4):2025-2033. DOI: 10.1109/ TDEI.2017.006164

[4] Du BX, Li A, Li J. Effects of AC and pulse voltage combination on surface charge accumulation and decay of epoxy resin. IEEE Transactions on Dielectrics & Electrical Insulation. 2016;23(4): 2368-2376. DOI: 10.1109/ TDEI.2016.7556515

[5] Lin C, Li C, He J, et al. Surface charge inversion algorithm based on bilateral surface potential measurements of cone- type spacer. IEEE Transactions on Dielectrics & Electrical Insulation. 2017; 24(3):1905-1912. DOI: 10.1109/ TDEI.2017.006496

[6] Du BX, Liang HC, Li J, et al. Temperature-dependent surface potential decay and flashover characteristics of epoxy/SiC composites. IEEE Transactions on Dielectrics & Electrical Insulation. 2018;25(2): 631-638. DOI: 10.1109/ TDEI.2017.006872

[7] Liang HC, Du BX, Li J, et al. Effects of nonlinear conductivity on charge trapping and de-trapping behaviors in epoxy/SiC composites under dc stress. IET Science Measurement & Technology. 2018;12(1):83-89. DOI: 10.1049/iet-smt.2016.0528

[8] Liang HC, Du BX, Li J, et al. Numerical simulation on the surface charge accumulation process of epoxy insulator under needle-plane corona discharge in air. IET Science Measurement & Technology. 2018; 12(1):9-16. DOI: 10.1049/iet- smt.2017.0164

[9] Vu-Cong T, Zavattoni L, Vinson P, et al. Surface charge measurements on epoxy spacer in HVDC GIS/ GIL in SF6. IEEE Electrical Insulation and Dielectric Phenomena. 2016. DOI: 10.1109/ CEIDP.2016.7785506

[10] Nakanishi K, Yoshioka A, Arahata Y, et al. Surface charging on epoxy spacer at DC stress in compressed SF$_6$ Gas. IEEE Power Engineering Review. 1983;PER-3(12):46-46. DOI: 10.1109/ MPER.1983.5520154

[11] Zhang B, Qi Z, Zhang G. Charge accumulation patterns on spacer surface in HVDC gas-insulated system: Dominant uniform charging and random charge speckles. IEEE Transactions on Dielectrics & Electrical Insulation. 2017;24(2):1229-1238. DOI: 10.1109/TDEI.2017.006067

[12] Zhang B, Qi Z, Zhang G. Thermal gradient effects on surface charge of HVDC spacer in gas insulated system. IEEE Electrical Insulation and Dielectric Phenomena. 2016. DOI: 10.1109/ CEIDP.2016.7785563

[13] Zhou HY, Ma GM, Li CR, et al. Impact of temperature on surface charges accumulation on insulators in SF6-filled DC-GIL. IEEE Transactions on Dielectrics & Electrical Insulation. 2017;24(1):601-610. DOI: 10.1109/ TDEI.2016.005838

[14] Sato S, Zaengl WS, Knecht A. A numerical analysis of accumulated surface charge on DC epoxy resin spaces. IEEE Transactions on Electrical Insulation. 2007;22(3):333-340. DOI: 10.1109/TEI.1987.298999

[15] Lutz B, Kindersberger J. Surface charge accumulation on cylindrical polymeric model insulators in air: Simulation and measurement. IEEE Transactions on

Dielectrics & Electrical Insulation. 2012;18(6):2040-2048. DOI: 10.1109/TDEI.2011.6118642

[16] Winter A, Kindersberger J. Stationary resistive field distribution along epoxy resin insulators in air under DC voltage. IEEE Transactions on Dielectrics & Electrical Insulation. 2012; 19(5):1732-1739. DOI: 10.1109/ TDEI.2012.6311522

[17] Winter A, Kindersberger J. Transient field distribution in gas-solid insulation systems under DC voltages. IEEE Transactions on Dielectrics & Electrical Insulation. 2014;21(1): 116-128. DOI: 10.1109/ TDEI.2013.004110

[18] He W, Liu XH, Yang F, et al. Numerical simulation of direct current glow discharge in air with experimental validation. Japanese Journal of Applied Physics. 2012;51(51):6001. DOI: 10.1143/JJAP.51.026001

[19] Hagelaar GJM. Solving the Boltzmann equation to obtain electron transport coefficients and rate coefficients for fluid models. Plasma Sources Science & Technology. 2005; 14(4):722-733. DOI: 10.1088/0963-0252/14/4/011

[20] Christophorou LG, Brunt RJV. SF$_6$// N$_2$ mixtures: Basic and HV insulation properties. IEEE Transactions on Dielectrics & Electrical Insulation. 2002;2(5):952-1003. DOI: 10.1109/94.469988

[21] Itikawa Y. Cross sections for electron collisions with nitrogen molecules. Journal of Physical & Chemical Reference Data. 2006;35(1): 31-53. DOI: 10.1063/1.1937426

[22] Gulley RJ, Cho H, Buckman SJ. The total elastic cross section for electron scattering from SF$_6$. Journal of Physics B: Atomic, Molecular and Optical Physics. 2000;33(8):L309-L315. DOI: 10.1088/0953-4075/33/8/105

[23] Phelps AV, Van Brunt RJ. Electron- transport, ionization, attachment, and dissociation coefficients in SF6 and its mixtures. Journal of Applied Physics. 1988;64(9):4269-4277. DOI: 10.1063/1.341300

[24] Zhou TC, Chen G, Liao RJ, et al. Charge trapping and detrapping in polymeric materials. Journal of Applied Physics. 2009;106(12):644-637. DOI: 10.1063/1.3273491

[25] Chen G, Zhao J, Zhuang Y. Numerical modeling of surface potential decay of corona charged polymeric material. IEEE International Conference on Solid Dielectrics. 2010:1-4. DOI: 10.1109/ ICSD.2010.5567997

[26] Min DM, Li ST, Li GC. The effect of charge recombination on surface potential decay crossover characteristics of LDPE. In: International Symposium on Electrical Insulating Materials. 2014. pp. 104-107. DOI: 10.1109/ISEIM.2014. 6870731

[27] Straumann U, Schuller M, Franck CM. Theoretical investigation of HVDC disc spacer charging in SF6 gas insulated systems. IEEE Transactions on Dielec-trics & Electrical Insulation. 2013; 19(6):2196-2205. DOI: 10.1109/ TDEI.2012.6396980

[28] Wang Q, Wang H, Peng Z, et al. 3-D coupled electromagnetic-fluid- thermal analysis of epoxy impregnated paper converter transformer bushings. IEEE Transactions on Dielectrics & Electrical Insulation. 2017;24(1): 630-638. DOI: 10.1109/TDEI.2016.005641

10

Analysis of Compensated Six-Phase Self-Excited Induction Generator Using Double Mixed State-Space Variable Dynamic Model

Kiran Singh

Abstract

In this article, a mixed current-flux d-q modeling of a saturated compensated six-phase self-excited induction generator (SP-SEIG) is adopted during the analysis. Modeling equations include two independent variables namely stator current and magnetizing flux rather than single independent variables either current or flux. Mixed modeling with stator current and magnetizing flux is simple by having only four saturation elements and beneficial in study of both stator and rotor parameters. Performance equations for the given machine utilize the steady-state saturated magnetizing inductance (L_m) and dynamic inductance (L). Validation of the analytical approach was in good agreement along with three-phase resistive or resistive-inductive loading and also determined the relevant improvement in voltage regulation of machine using series capacitor compensation schemes.

Keywords: mixed double state space variables, self-excitation, six-phase, compensation, induction generator, non-conventional energy

Introduction

Traditionally, synchronous generators have been used for power generation, but induction generators are increasingly being used these days because of their relative advantageous features over conventional synchronous generators. The need for external reactive power limits the application of an induction generator as isolated unit. The use of SEIG, due to its reduced unit cost, simplicity in operation and ease of maintenance are most suited in such system. These entire features facilitate the operation of induction generator in stand-alone mode to supply far-flung areas where extension of grid is economically not viable. The stand-alone SEIG can be used with conventional as well as non-conventional energy sources to feed remote single family, village community, etc. in order to expedite the electrification of rural and remote locations. A detailed dynamic performance of induction generator 'IG' operating in different modes i.e. isolated and grid-connected is necessary for the optimum utilization of its various favorable features. The investigations spread over last two decades also indicate the technical and economic vitality of using number of phases higher than three in AC machines for applications in marine ships, thermal power plant to drive induced draft fans, electric vehicles and circulation pumps in nuclear power plants etc. In this area, research is still in its early stage, yet some extremely great authority's findings have been reported in the previous literatures indicating the general expediency of multi-phase systems. The literature regarding multi-phase IG is nearly not available since it has only three findings before 2004.

The first article on multi-phase induction generator is appeared in 2005, along with rest of theoretical and practical works on SP-SEIG using single state- space variables either stator and rotor d-q axis currents or stator and rotor d-q axis flux linkages so far as reported [1]. On the basis of previous article reviews, before companion paper of 2015 [1], there were no literatures on modeling and analysis of SP-SEIG, using d-q axis components of stator current and magnetizing flux as mixed state–space variables. In the view of novelty, such mathematical modeling and analysis were carried out in detail for three-phase SEIG only, in a very few available literatures and some of which are mentioned by [2]. The purpose of this article is also to accomplish a similar task for modeling and analysis of SP-SEIG with series compensation scheme using double state-space variables as were proposed by companion paper without compensation. The simulation is performed on series compensated SP-SEIG by using 4th order Runge-Kutta subroutine in Matlab software.

Modeling description

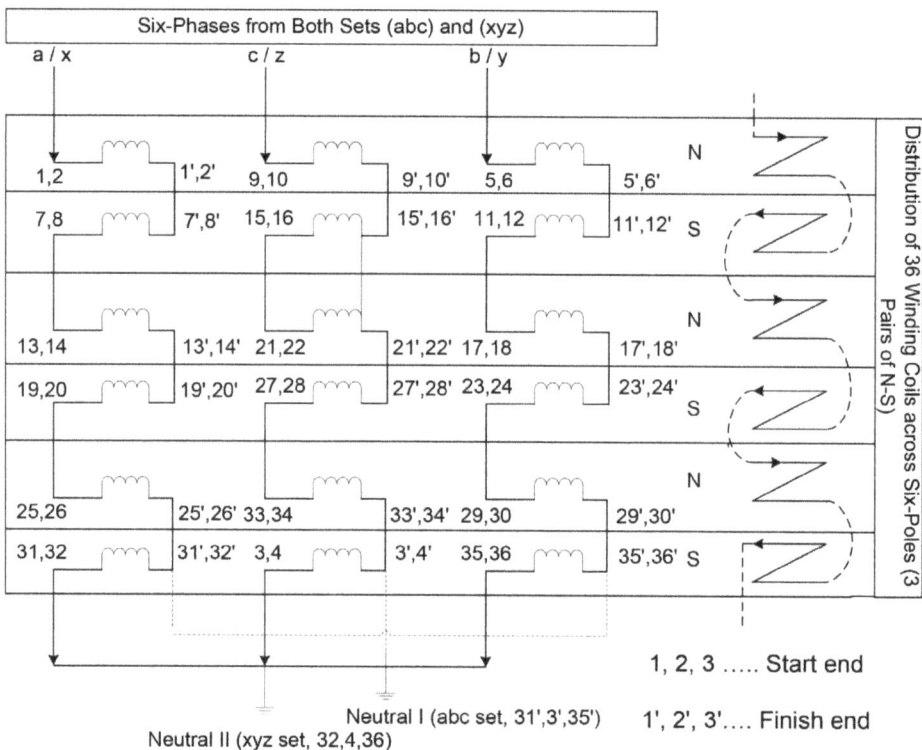

Figure 1. *Distribution of 6-phases in 36 slots of 6 pole induction machine.*

Concerns about mathematical modeling of SP-SEIG, short-shunt series compensation capacitors and static resistive 'R' and balanced three-phase reactive 'R-L' loads as previously discussed in [1, 3] are not described in this section, only briefly summarized along with newly addition of long-shunt series compensation capacitors.

SP-SEIG model

A basic two-pole, six-phase induction machine is schematically described by its stator and rotor axis [1]. In which, six stator phases of both sets, *a, b, c* and *x, y, z* (set I and II, respectively) are arranged to form two sets of uniformly distributed star configuration, displaced by an arbitrary angle of 30 electrical degree gravitate asymmetrical winding structure. The distribution of 6-phases in 36 slots of 6 pole induction machine is also shown in **Figure 1**. Previously, voltage equations using single state-space variable namely flux linkage were used in the expanded form for six-phase induction machine.

After simplification of voltage equations using double mixed state-space variables namely stator currents and magnetizing fluxes in the machine model following form occurs from Eqs. (1)–(39) of [1]:

$$[V_{dq}] = [H] [dX_{dq}]/dt + [J][X_{dq}] \tag{1}$$

where $[V_{dq}] = [V_{d1}\ V_{q1}\ V_{d2}\ V_{q2}\ 0\ 0]^t$, $[X_{dq}] = [i_{d1}\ i_{q1}\ i_{d2}\ i_{q2}\ \psi_{dm}\ \psi_{qm}]^t$ and matrices [H] and [J] are given by Eqs. (2) and (3), respectively.

$$7 \begin{bmatrix} -L_{\sigma 1} & 0 & 0 & 0 & 1 & 0 \\ 0 & -L_{\sigma 1} & 0 & 0 & 0 & 1 \\ 0 & 0 & -L_{\sigma 2} & 0 & 1 & 0 \\ 0 & 0 & 0 & -L_{\sigma 2} & 0 & 1 \\ L'_{\sigma r} & 0 & L'_{\sigma r} & 0 & \left(1 + \dfrac{L'_{\sigma r}}{L'_{dd}}\right) & \dfrac{L'_{\sigma r}}{L_{dq}} \\ 0 & L'_{\sigma r} & 0 & L'_{\sigma r} & \dfrac{L'_{\sigma r}}{L_{dq}} & \left(1 + \dfrac{L'_{\sigma r}}{L_{qq}}\right) \end{bmatrix} \tag{2}$$

$$\begin{bmatrix} -r_1 & wL_{\sigma 1} & 0 & 0 & 0 & -w \\ -wL_{\sigma 1} & -r_1 & 0 & 0 & w & 0 \\ 0 & 0 & -r_2 & wL_{\sigma 2} & 0 & -w \\ 0 & 0 & -wL_{\sigma 2} & -r_2 & w & 0 \\ r'_r & -(w-w_r)L'_{\sigma r} & r'_r & -(w-w_r)L'_{\sigma r} & \dfrac{r'_r}{L_m} & -(w-w_r)\left(1+\dfrac{L'_{\sigma r}}{L_m}\right) \\ (w-w_r)L'_{\sigma r} & r'_r & (w-w_r)L'_{\sigma r} & r'_r & -(w-w_r)\left(1+\dfrac{L'_{\sigma r}}{L_m}\right) & \dfrac{r'_r}{L_m} \end{bmatrix} \tag{3}$$

The nonlinear equations of voltage and current across the shunt excitation capacitor and series compensation capacitors (short-shunt and long-shunt) can be transformed into d-q axis by using reference frame theory, i.e. Park's (d_{q0}) transformation [4], are given by Section 2.2 of [1] and (Section 2.3 of [1] and by following Section 2.2), respectively. Modeling of static loads is also given in Section 3 of [1].

Modeling of long-shunt capacitors

Current through series capacitors Cls1 and Cls2 (in case of long shunt), connected in series with winding set I and II, respectively, is same as the machine current. The machine current along with series capacitance determine the voltage across series long-shunt capacitor and when transformed in to d-q axis by using Park's transformation is given in Eqs. (4) and (5) [5–7].

Table 1. *Machine model symbols.*

$$\rho V_{q1ls} = i_{q1}/C_{ls1}$$
$$\rho V_{d1ls} = i_{d1}/C_{ls1}$$
$$\rho V_{q2ls} = i_{q2}/C_{ls2}$$
$$\rho V_{d2ls} = i_{d2}/C_{ls2}$$

(4)

and the load terminal voltage is expressed as

$$V_{Lq1} = V_{q1} + V_{q1ls}$$
$$V_{Ld1} = V_{d1} + V_{d1ls}$$
$$V_{Lq2} = V_{q2} + V_{q2ls}$$
$$V_{Ld2} = V_{d2} + V_{d2ls}$$

(5)

The remaining symbols of machine model have their usual meanings from Ref. [1] and **Table 1**.

Methodology

In this section, a numerical method is introduced to the solution of Eqs. (1)–(3); where double mixed current flux state space model is discussed by [1]. The ordinary linear differential equations can be solved by the analytical technique rather than approximation method. Eq. (1) is non-linear differential and cannot be solved exactly with high expectations, only approximations are estimated numerically by computer technique using 4th order Runge-Kutta method or classical Runge-Kutta method or often referred as "RK4" as so commonly used [8]. The analytical response of compensated SP-SEIG in only single operating mode is carried out under significant configuration using RK4 subroutine implemented in Matlab Mfile. The dynamic performances were determined under no load, R load and R-L loading condition in only the single mode of excitation capacitor bank, and in both modes of compensating series capacitor bank. The following analytical dynamic responses of series compensated SP-SEIG is considered for the validity of proposed approaches in this chapter.

- During R and R-L loading with short shunt compensation.

- During R and R-L loading with long shunt compensation.

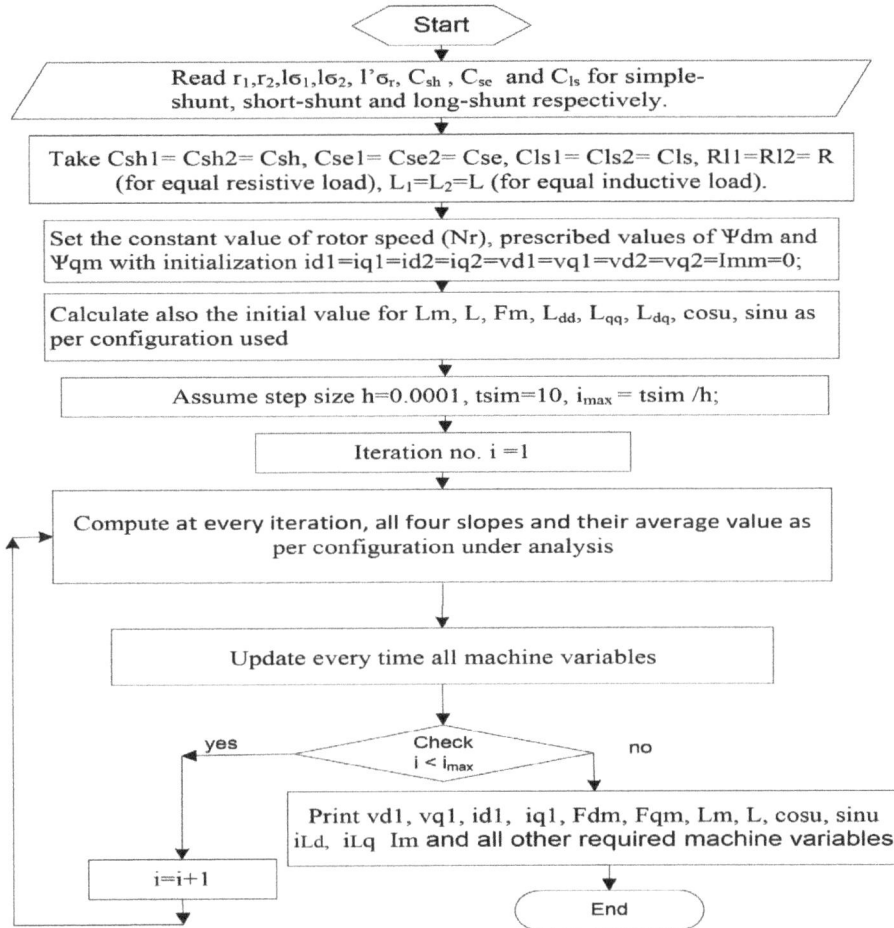

Figure 2. *Algorithm for Runge-Kutta method implemented for SP-SEIG under constant rated speed at 1000 RPM.*

Both analytical responses are well detailed in Section 4. The analytical study of compensated SP-SEIG is given in Section 4 by using an explicit MATLAB program incorporates the RK4 method. An Algorithm of RK4 method for the analysis of compensated SP-SEIG is also shown in **Figure 2**. The parameters of studied machine and saturation dependent coefficients of system matrix [H] are also reported by [1] for further dynamic analysis of saturated compensated SP-SEIG using double mixed state space variable model under constant rotor speed along with appropriate initial estimated variables values which are also responsible in the development of rated machine terminal voltage and it depends on other machine variables.

Analytical response

Analytical performances of compensated SP-SEIG are illustrated in **Figures 3–6** along with sudden switching of R load of 200 Ω and a balanced three-phase R-L load (200 Ω in series with 500 mH) at t = 2 s when short-shunt and long-shunt compensation along both three-phase winding sets. These waveforms are load, generated voltage and current amplitude has been dropped few more volts and amperes, respectively, compared to R load. Combined correspond to d-q axis stator voltage, d-q axis stator current, d-q axis load current, d-q axis magnetizing flux, d-q axis series capacitor voltage and magnetizing current. In the case of R-L amplitude of magnetizing flux, steady-state saturated magnetizing inductance and dynamic (tangent slope) inductance along with the combined amplitude waveform of d-axis stator current and load current during no-load and sudden switching of R load of 200 Ω at t = 2 s are also shown in **Figures 3–6**. As per discussion of [1], generated RMS value of steady state terminal voltage and current is about 225 V line to line and 4.46 A at rated speed of 1000 RPM with the value of excitation capacitance of 38.5 μF per phase in simple-shunt configuration for the selected machine in this article.

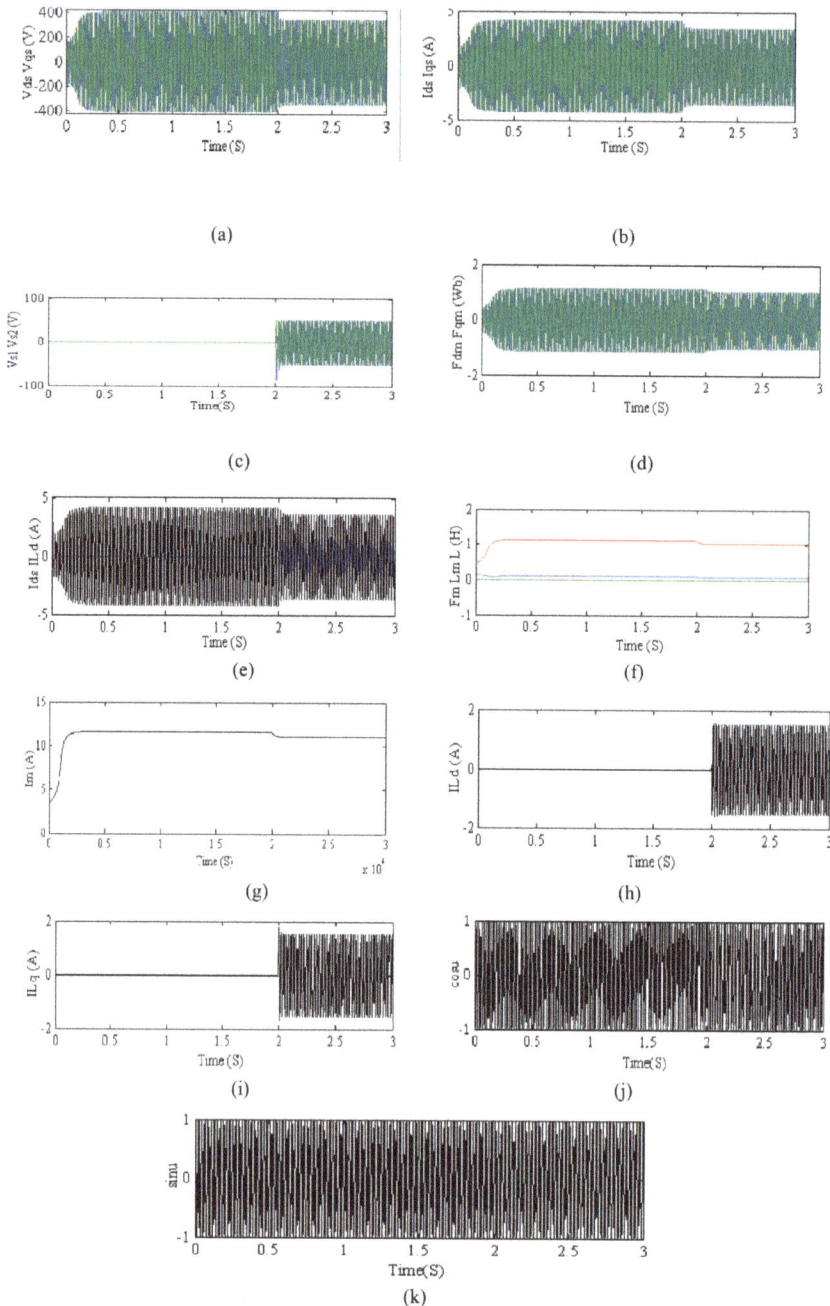

Figure 3. *Analytical waveforms during sudden switching of R load of 200 Ω at t = 2 s.*

In proposing stator current and magnetizing flux mixed variable model, it has been observed that SP-SEIG terminal voltage and current build-up from their initial values to final steady state values entirely depends upon its initial few Weber values of d-and q-axis magnetizing flux at constant rated speed of 1000 RPM.

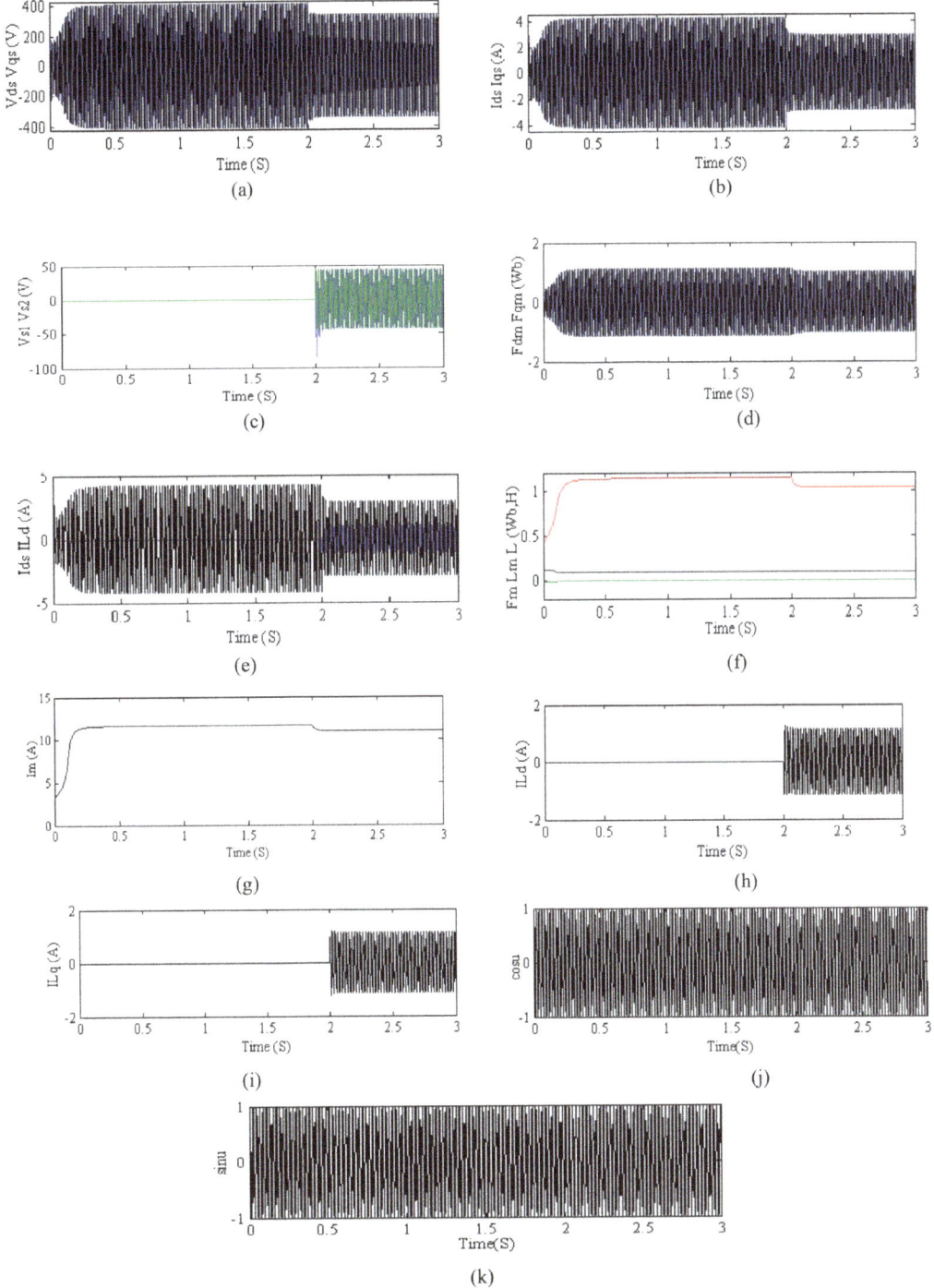

Figure 4. *Analytical waveforms during sudden switching of RL load at t = 2 s.*

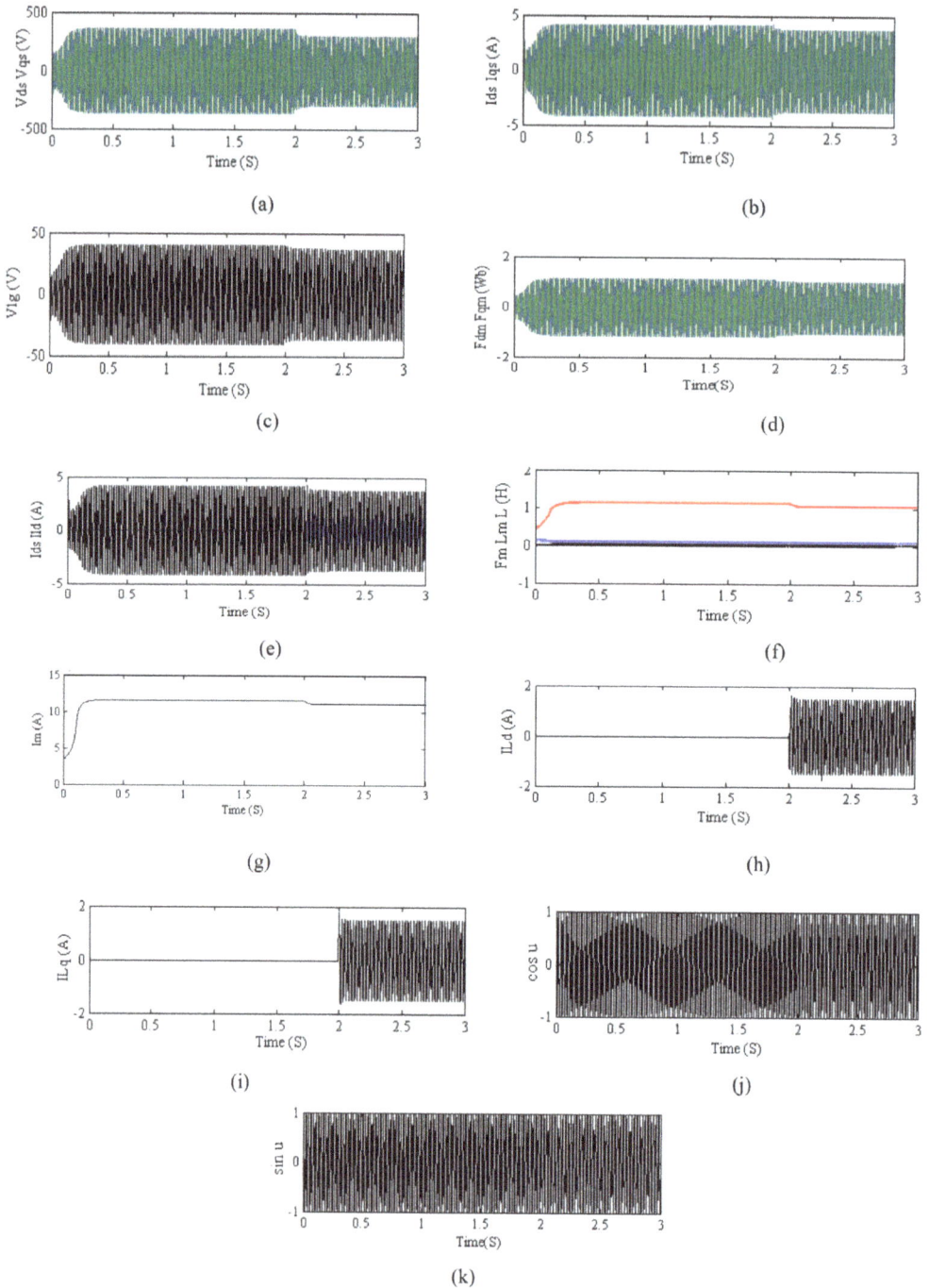

Figure 5. *Analytical waveforms during sudden switching of R load of 200 Ω at t = 2 s.*

Short-shunt series compensated SP-SEIG

Performance of short-shunt SP-SEIG in only its single mode of operation using the values of excitation and series capacitor banks of 38.5 and 108 μF per phase, respectively, has been predicted from the built explicit MATLAB program using RK4 subroutine. Computed waveforms are shown in **Figures 3** and **4**. Application of short-shunt scheme results in overvoltage across the generator terminals as shown in **Figure 3a**, the per phase voltage level is more than the voltage level of **Figure 5a** and it is illustrated in **Figure 3c**.

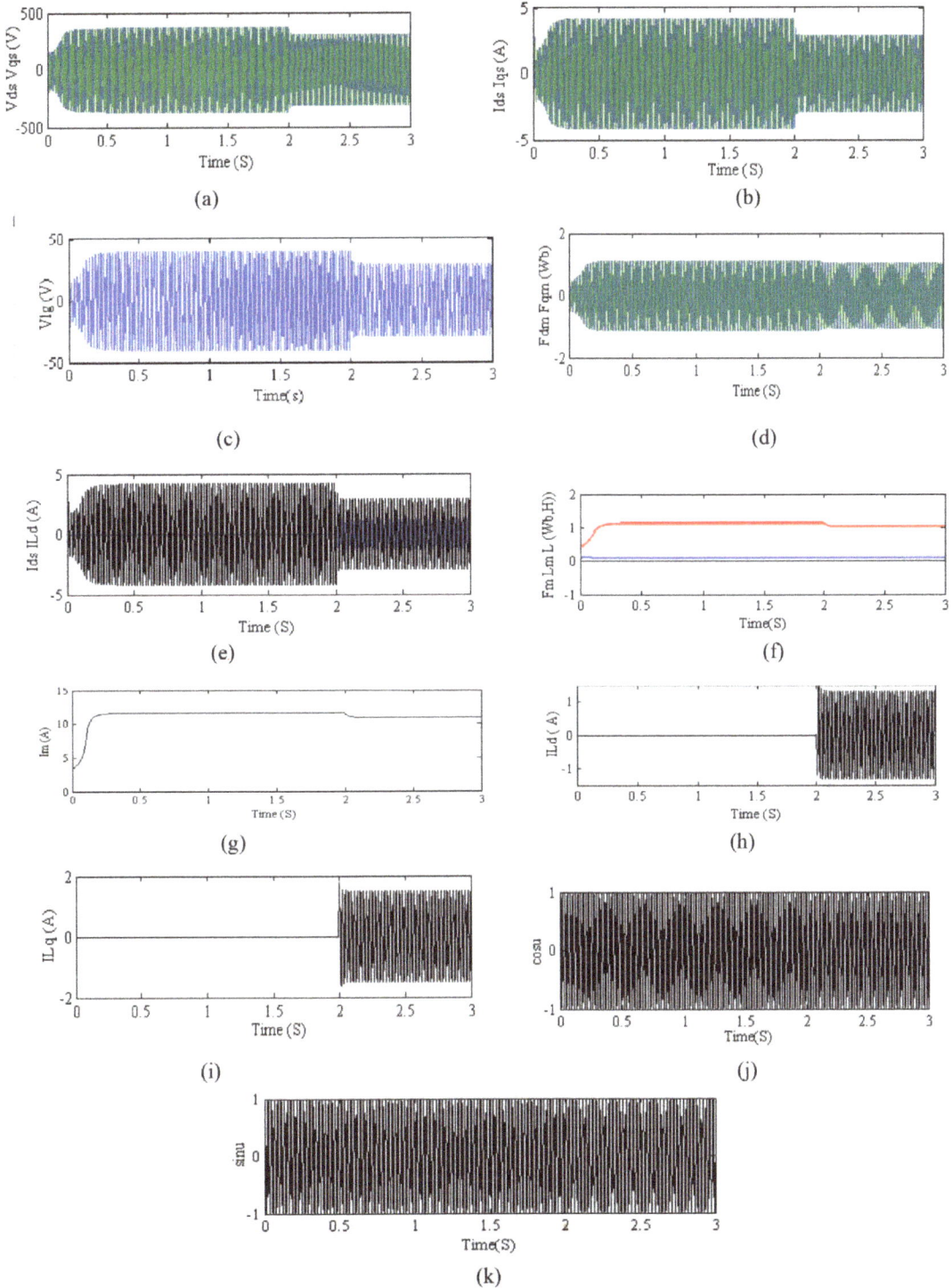

Figure 6. *Analytical waveforms during sudden switching of R-L load at t = 2 s.*

When both three-phase winding sets are connected in short-shunt configuration with independent R loading

The analytical d-q waveform of voltage, current, magnetizing flux and magnetizing current during no-load and sudden switching of R load of 200 Ω at t = 2 s are shown in corresponding **Figure 3a**, **b, d** and **g**. In addition, combined amplitude waveforms of magnetizing flux, steady-state saturated magnetizing inductance and dynamic (tangent slope) inductance is shown in **Figure 3f**. The angular displacements of the magnetizing current space vector with respect to the d-axis of the common reference frame are also shown in **Figures 3j** and **k**. The combined d-q axis voltage drop across the series short-shunt capacitor is given in **Figure 3c** and combined amplitude waveform of d-axis stator current and load current during noload and sudden switching of R load of 200 Ω at t = 2 s is given in **Figure 3e**. The d- and q-axis load currents are also depicted in **Figure 3h** and **i**, respectively, during sudden switching of R load of 200 Ω at t = 2 s.

When both three-phase winding sets are connected in short-shunt configuration with independent R-L loading

In the same order of figure numbers, all computed waveforms are shown in **Figure 4** with sudden switching of R-L load (200 Ω resistance in series with 500 mH inductor) at t = 2 s. Analytical generated RMS steady state voltage and corresponding current at rated speed of 1000 RPM are shown in **Figure 4a** and **b**. Drops in d-q-axis generating and lagging load currents are shown in **Figure 4b**, **h** and **i**.

Long-shunt series compensated SP-SEIG

It is also seen that like short-shunt compensation, long-shunt compensation is also self- regulating in nature. In the same manner, analysis of long-shunt SP-SEIG along with a single mode of excitation and the series capacitor banks of 38.5 and 350 F respectively, have also been computed and predicted by using the RK4 subroutine and illustrated in **Figures 5** and **6.** Here, the value of series capacitor is more than the twice of short-shunt series capacitor.

When both three-phase winding sets are connected in long-shunt configuration with independent R loading

The analytical waveform of voltage, current, magnetizing flux and magnetizing current during no-load and sudden switching of R load of 200 Ω at t = 2 s with long- shunt compensation along both three-phase winding sets are respectively shown in **Figure 5a**, **b, d** and **g**. Combined amplitude waveforms of magnetizing flux, steady- state saturated magnetizing inductance and dynamic (tangent, slope) inductance is shown in **Figure 5f**. The angular displacements of the magnetizing current space vector with respect to the d-axis of the common reference frame are also shown in **Figure 5j** and **k**. The application of the long-shunt scheme results in less overvoltage or reduced terminal voltage across the generator terminals. As it is shown in **Figure 5a**, the per phase voltage level is less than the voltage level of **Figure 3a**. It gives evidence that long-shunt SP-SEIG is able to deliver output power at reduced terminal voltages, as shown in **Figure 5c**. The combined d-q axis voltage drop across the series short-shunt capacitor is given in **Figure 5c** and combined amplitude waveform of d-axis stator current and load current during no-load and sudden switching of resistive load of 200 Ω at t = 2 s is given in **Figure 5e**. The d- and q-axis load currents are also depicted in **Figure 5h** and **i** respectively, during sudden switching of resistive load of 200 Ω at t = 2 s. The steady state no-load voltage is generated at rated speed of 1000 RPM.

When both three-phase winding sets are connected in long-shunt configuration with independent R-L loading

In the same order of figure numbers, all computed waveforms are also shown in **Figure 6** with sudden switching of R-L load (200 Ω resistance in series with 500 mH inductor) at t = 2 s. The RMS analytical value of steady state voltage is generated at rated speed of 1000 RPM. Same like as Section 4.1, the d-q axis drops in generating and lagging load currents are also given in **Figure 6b**, **h** and **i**.

Discussion

In the machine model, two mixed state space variables have been chosen in place of single state space variable (general case). Double mixed stator current and air- gap flux state space model belongs to one of the more complex model types compared to remaining mixed variable model of stator flux linkages-stator current, rotor flux linkages-rotor current, rotor flux linkages-stator current, stator flux linkages-rotor current, air-gap flux-rotor current, magnetizing current-stator flux linkages and rotor flux linkages-magnetizing current. On the other perspective, considerably simpler than d-q axis winding current model and has simple matrix model. Short-shunt compensation results in overvoltage across the generator terminals during no-load. Whereas, long-shunt compensation gives reduced terminal voltage as compared to short-shunt compensation. In long-shunt configuration, deep saturation provides higher level of magnetizing (or stator) current against large load current at sudden switching of R (or RL) load. While, sometimes, in long-shunt compensation scheme, even reduced generated no load terminal voltage as compared to short-shunt scheme, can be capable of almost same generated total output power. In this fashion, when one moves from simple-shunt to short-shunt (or long-shunt) compensated SP-SEIG, voltage regulation has to be improved by maintaining almost similar magnitude of voltage response after load, as was in simple-shunt SP-SEIG at no-load in **Figure 5a** of Ref. [1]. Proposed saturated machine model will be applied in air-gap flux field orientation vector control strategy.

Conclusion

Mixed stator current and air-gap flux as a double state-space variables model preserves information about both stator and rotor parameters. A careful value selection of the combination, i.e. shunt and series capacitors may avoid the excessive voltage across the terminals of SP-SEIG during sudden switching of machine load. It is noticed that involvement of extra supplied reactive power as per self- regulating nature of short-shunt as compared to long-shunt series capacitors in each lines, retains the similar output profile of simple-shunt scheme, when machine load is suddenly switched on after few seconds. In both cases (R and R-L loading), the little bit of marked voltage drops were occurred during R-L loading compared to the R loading when variation from no load to full load.

Author details

Kiran Singh

Department of Electrical Engineering, Indian Institute of Technology, Roorkee, Uttarakhand, India

*Address all correspondence to: kiransinghiitr@gmail.com

References

[1] Singh K, Singh GK. Modelling and analysis of six-phase self-excited induction generator using mixed stator current and magnetizing flux as state- space variables. Electric Power Components and Systems. 2015;43: 2288-2296

[2] Liao YW, Levi E. Modelling and simulation of a stand-alone induction generator with rotor flux oriented control. Electric Power Systems Research. 1998;46:141-152

[3] Singh K, Singh GK. Stability assessment of isolated six-phase induction generator feeding static loads. Turkish Journal of Electrical Engineering & Computer Sciences. 2016;24:4218-4230

[4] Krause PC. Analysis of Electric Machinery. New York: McGraw Hill Book Company; 1986

[5] Khan MF, Khan MR. Generalized model for investigating the attributes of a six-phase self-excited induction generator over a three-phase variant. International Transactions on Electrical Energy Systems. 2018;28:e2600

[6] Khan MF, Khan MR. Performance analysis of a three phase self-excited induction generator operating with short shunt and long shunt connections. IEEE Biennial International Conference on Power and Energy Systems: Towards Sustainable Energy (PESTSE). 2016:1-6

[7] Chermiti D, Abid N, Khedher A. Voltage regulation approach to a self- excited induction generator: Theoretical study and experimental validation. International Transactions on Electrical Energy Systems. 2017;27(5):e2311

[8] Kohler Werner E, Johnson Lee W. Elementary Differential Equations with Boundary Value Problems. 2nd ed. Addison-Wesley; 2006

11

Five-Phase Line Commutated Rectifiers

Mahmoud I. Masoud

Abstract

Multiphase systems including multiphase generators or motors, especially five- phase, offer improved performance compared to three-phase counterpart. Five- phase generators could generate power in applications such as, but not limited to, wind power generation, electric vehicles, aerospace, and oil and gas. The five-phase generator output requires converter system such as AC-DC converters. This chapter introduces the basic construction and performance analysis for uncontrolled/controlled five-phase line commutated rectifier guided by numerical examples. This rectifier is suitable for wind energy applications to be the intermediate stage between five-phase generator and DC load or inverter stage. The filtration for AC side and effects of source inductances are detailed in other references. Here, this chapter gives the reader a quick idea about the analysis and performance of multiphase line commutated rectifiers and specifically five phase.

Keywords: five phase, multiphase generators, wind turbine, controlled and uncontrolled rectifiers, AC-DC converters, line commutated rectifiers

Introduction

Nowadays, energy generation using renewable energy sources is considered as contemporary issue. Many countries are experiencing economic pressures because of the reduction in oil prices. Moreover, energy that is produced by burning fossil fuels is mainly responsible for the global air pollution problem, and consequently, it is not environmentally friendly. As a consequence, many countries are impelled to concentrate on renewable energy sources including generation, management, and planning. Renewable energy share, of the total energy consumption by the end of 2016, is 24.5% where it was 3.3% in 2010 [1]. This number is tremendously increasing each year. Accordingly, most research institutes foster research in the energy sector to go to clean and green energy generation. From renewable energy sources, wind energy is considered to be a pillar using appropriate generator. By the end of 2016, 29 countries have more than 1 GW in operation from wind energy, and the total capacity generated for the world is 487 GW. Only in 2016 a 23.4 GW is added due to wind energy generation [1]. Incontrovertibly, wind power becomes a big industry and a big topic for research. It includes mechanical parts and electrical parts. Accordingly, a generator that converts wind mechanical power to electrical power is a must. But, the power generated using wind turbines is not constant and could fluctuate. Also, the load profile fluctuates with respect to time. To accomplish this, specially designed power systems for meeting these demands are built which can feed these drastic load changes. In all cases, the generator output voltage should be rectified. Owing to the numerous advantages when compared to its three-phase counterpart such as high fault-tolerant capability, five-phase direct-drive permanent magnet generators are a key area of focus in power generation with renewable and wind energy systems [2–6]. Hence, the generator output requires a converter, such as an AC-DC rectifier, to match load requirements. Rectifiers are broadly classified into two types, namely, line-commutated rectifiers (LCRs) and power factor control rectifiers. LCRs are either uncontrolled (using diodes) or controlled (using thyristors). This type of rectifiers operates on power frequency or generated frequency and has the advantages of simple construction and low cost. In a controlled LCR, the DC-link voltage can be controlled by adjusting the firing angle, thus facilitating high power capability, easy control, and a simple gating system [7, 8]. However, LCRs still have a low input power factor, high input current total harmonic distortion (THD), and unidirectional current flow [9, 10]. To address such problems, power factor control rectifiers, which operate at a high switching frequency, such as pulse-width modulated (PWM) rectifiers are used where the filter size for PWM rectifiers is smaller owing to the higher switching frequencies [9, 11, 12]. However, one of the main drawbacks of using PWM rectifiers is the limitation of the switch current and voltage, which restricts the use of such rectifiers in high-power-scale applications [3, 5, 13, 14].

This chapter introduces uncontrolled/controlled line-commutated five-phase rectifier as we target high-power wind energy applications. The study includes the analysis of both AC side and DC side. Moreover, the gating signal generation is introduced. The study includes ideal source to represent the PM generator. Effect of source inductance and filtration of AC side are introduced in Refs. [3, 5].

Five-phase source

The phase voltage of the five-phase source, which is the output of the five-phase generator, can be written as

$$v_{jn} = V_{ph\_max} \sin \left(\omega t - \frac{2\pi k}{5} \right) \tag{1}$$

where j represents phases a, b, c, d, and e, respectively, and k equals 0, 1, 2, 3, and 4, respectively. The five-phase source has two line voltages (adjacent and nonadjacent) [3, 5]. The line voltages can be calculated using phasor diagram shown in **Figure 1**.

The line voltage for adjacent phases has amplitude 1.1756 V_{ph_max} and leads to the phase voltage by 54°; see **Figure 1a**. The line voltage for nonadjacent phases has amplitude 1.902 V_{ph_max} and leads to the phase voltage by 18°; see **Figure 1b**.

Example 1: *For a 200 V (peak value-phase voltage) five-phase source, calculate and plot the phase and line voltage waveforms.*

Solution: Phase voltages are

$$v_a = 200 \sin \ (\omega t), \quad v_b = 200 \sin \ \left(\omega t - \frac{2\pi}{5} \right), \quad v_c = 200 \sin \ \left(\omega t - \frac{4\pi}{5} \right),$$

$$v_d = 200 \sin \ \left(\omega t + \frac{4\pi}{5} \right), \quad v_e = 200 \sin \ \left(\omega t + \frac{2\pi}{5} \right) \quad V$$

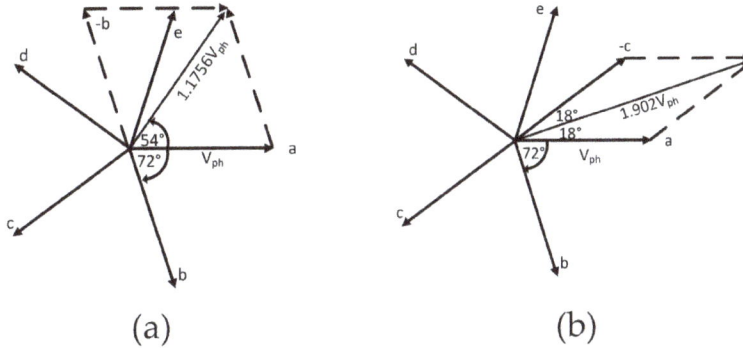

Figure 1. *Phasor diagram for five-phase source: (a) line voltage for adjacent phases and (b) line voltage for nonadjacent phases.*

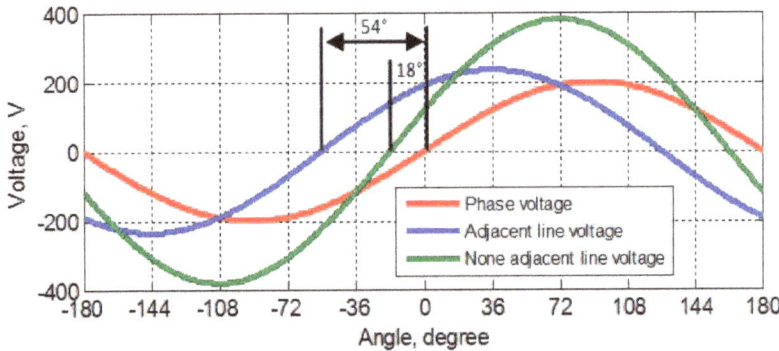

Figure 2. *Phase voltage, adjacent line voltage, and non-adjacent line voltage.*

Figure 2 shows the line voltages for adjacent and non-adjacent phase voltages. The line voltage for adjacent phases is

$$v_{ac} = 200^* 1.902 \sin \ \left(\omega t + \frac{\pi}{10} \right) = 380.4 \sin \ \left(\omega t + \frac{\pi}{10} \right) \quad V$$

The line voltage for nonadjacent phases is

$$v_{ac} = 200^* 1.902 \sin \ \left(\omega t + \frac{\pi}{10} \right) = 380.4 \sin \ \left(\omega t + \frac{\pi}{10} \right) \quad V$$

Uncontrolled five-phase rectifier

The five-phase uncontrolled rectifier is shown in **Figure 3**. It consists of 10 diodes where the load is fed from five-phase half wave connection, five switches *D1–D5*, and return path being via another half wave connection, five switches *D6–D10*, to one of the five supply lines.

The load voltage or output DC voltage can be calculated by the addition of the two five-phase half-wave voltages, relative to the supply neutral point n, appearing at the positive and negative sides of the load, respectively. The voltage is 10 pulses in nature having its maximum instantaneous value of that of line voltages for nonadjacent phases and leads the phase voltage by $\pi/10$. The load voltage follows in turn 10 sinusoidal voltages during one cycle, those being v_{ac}, v_{ad}, v_{bd}, v_{be}, v_{ce}, v_{ca}, v_{da}, v_{db}, v_{eb}, and v_{ec}. The average voltage can be calculated with the aid of **Figure 4**. The repeated period accommodates 36° and the limits are -18° to +18°.

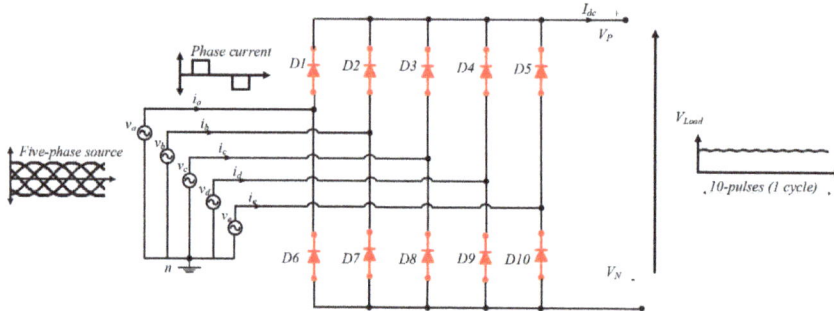

Figure 3. *Five-phase uncontrolled line commutated rectifier.*

The average or mean load voltage can be calculated as

$$V_{mean} = \frac{1}{\pi/5}\left[\int_{-\frac{\pi}{10}}^{\frac{\pi}{10}} 1.902 V_{ph\_max}\cos\omega t\, d\omega t\right] \tag{2}$$

$$V_{mean} = 1.87\, V_{ph\_max} \tag{3}$$

The waveform of supply voltage (phase voltage), positive voltage with respect to neutral (V_p), negative voltage with respect to neutral (V_N), load voltage, diode 1 voltage, diode 1 current, and supply current are shown in **Figure 5** parts 'a' to 'e'. The peak reverse voltage for one diode is the line voltage of nonadjacent phases and as indicated in **Figure 5c**.

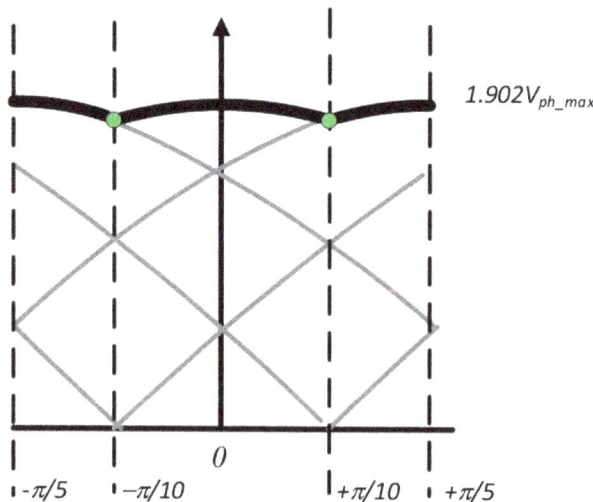

Figure 4. *The load voltage limits.*

The ripple voltage for DC side is

$$\% \ Voltage \ ripple = \frac{Maximum \ value - Minimum \ value}{V_{mean}} * 100 \qquad (4)$$

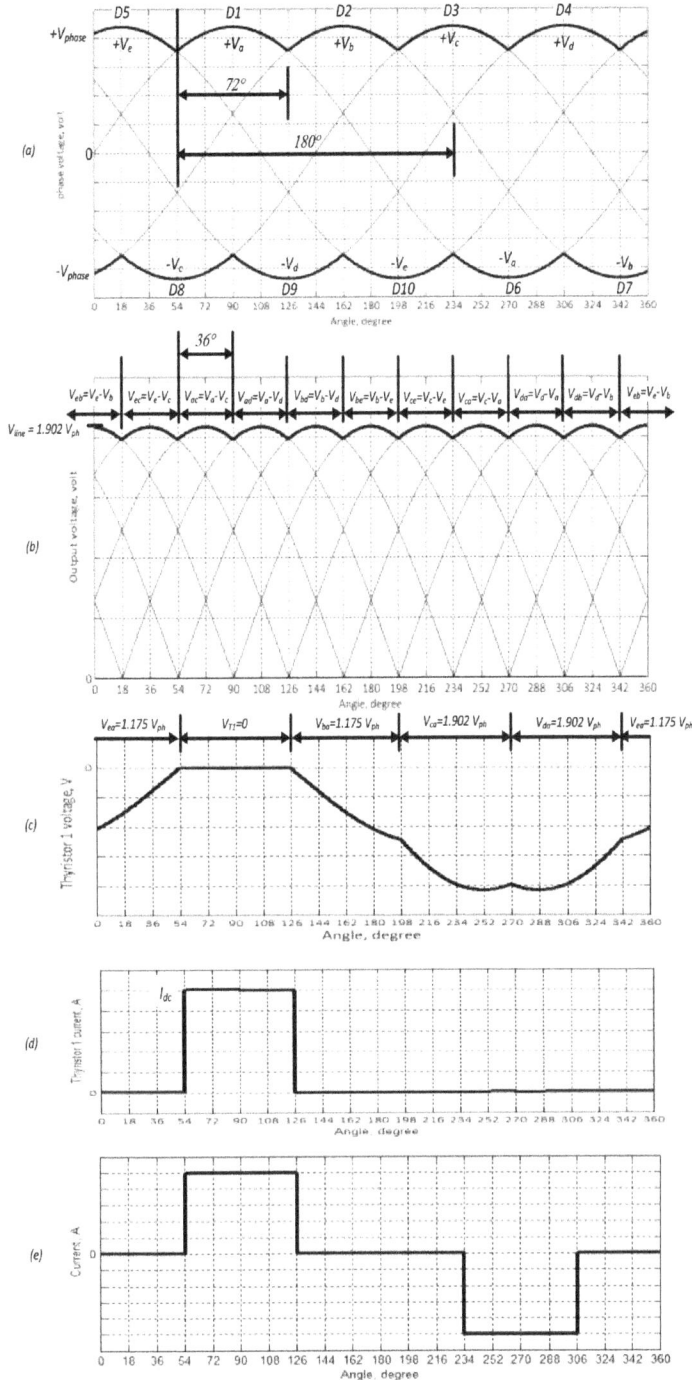

Figure 5. *Five-phase uncontrolled rectifier waveforms: (a) supply voltage, positive voltage with respect to neutral VPN, and negative voltage with respect to neutral VNN, (b) output voltage waveform, (c) diode 1 current, (d) diode 1 current, and (e) supply current for phase 'a'.*

The DC-side harmonics are generated at $10kf$, where f is the fundamental frequency of the AC source or generated power frequency and k equals (1, 2, 3, etc.). The lowest order harmonic represents 2.1% of the average DC value as shown in **Figure 6**. The supply current is a quasi-square wave in its nature as shown in **Figure 4e**. The load current is constant as the load is considered highly inductive. Switch current operates for only $2\pi/5$ from one cycle; accordingly, the switch average current can be calculated as in Eq. (5) which represents 20% of the DC-link current.

$$I_{av\_switch} = \frac{1}{2\pi} \left[\int_{\frac{3\pi}{10}}^{\frac{7\pi}{10}} I_{dc} \ d\omega t \right] = 0.2 \ I_{dc} \tag{5}$$

This means five-phase rectifier could have switches with lower ratings than three-phase counterpart which feed the same load as the average value of switch current in three-phase system is 33.3% [8, 15]. Part 'e' in **Figure 5** shows that supply current is a quasi-square wave where input current comprises $2\pi/5$ alternating polarity blocks, with each phase displaced to others by $2\pi/5$. The input phase current can be expressed as [5, 8, 11, 15].

$$i_s(t) = \sum_{n=1, 3, 5, \dots}^{\infty} \frac{4I_{dc}}{n\pi} \ \sin \ \frac{n\pi}{5} \ \sin \ (n\omega t) \tag{6}$$

Substituting $n = 5$ (or any odd multiple of 5) into Eq. (6) results in $i_{s5}(t) = 0$. The fundamental input current ($n = 1$) is

$$i_{s1}(t) = 0.7484 \ I_{dc} \ \sin (\omega t) \tag{7}$$

and rms value of fundamental current is

$$I_{s1} = 0.529 \ I_{dc} \tag{8}$$

The rms value of supply current is calculated as

$$I_s = \sqrt{\frac{1}{\pi} \int_{3\pi/10}^{7\pi/10} I_{dc}^2 \, d\omega t} = \sqrt{\frac{2}{5}} I_{dc} = 0.6324 I_{dc} \tag{9}$$

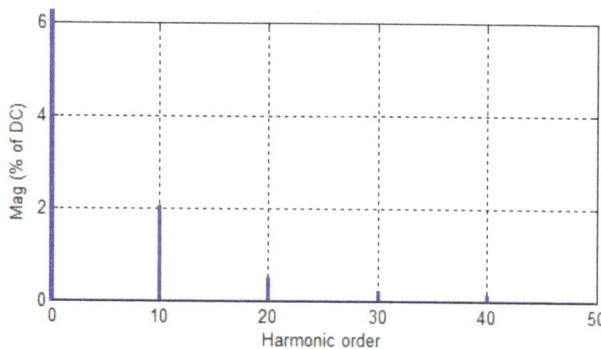

Figure 6. *Harmonic spectrum for load voltage percentage of average DC value.*

The harmonic contents of the supply current normalized to the fundamental component (I_{sn}/I_{s1}) are shown in **Figure 7**. The harmonic factor (HF_n) is defined as

$$HF_n\% \ = \ \frac{I_{sn}}{I_{s1}}*100 \tag{10}$$

Example 2: *A 240 V, 50 Hz, five-phase uncontrolled rectifier is connected to highly inductive load and takes 25 A. Calculate:*

a.   *Average load voltage*

b.   *The value of lowest order harmonic in DC side*

c.   *Maximum and minimum DC side voltage and voltage ripple*

d.   *The root mean square value of supply current*

e.   *The root mean square value of the supply current fundamental*

f.   *The third, fifth, and seventh harmonic factors for supply current*

Solution: **Figure 8** shows the load voltage and supply voltage for the uncontrolled rectifier when the supply voltage has phase and rms value of 240 V.

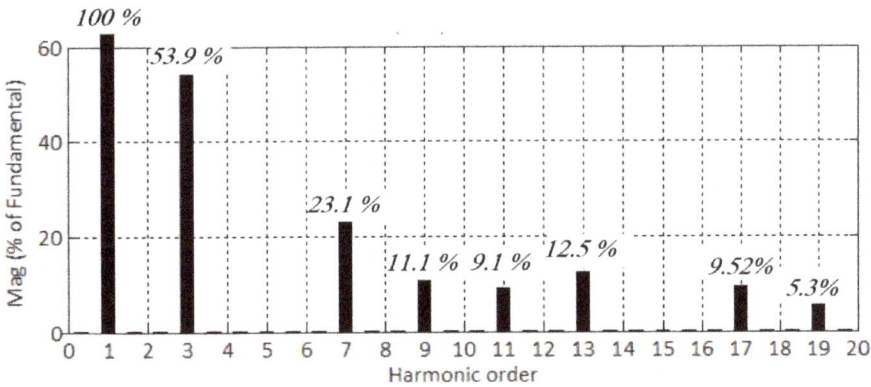

Figure 7. *Supply current spectrum normalized to the fundamental component.*

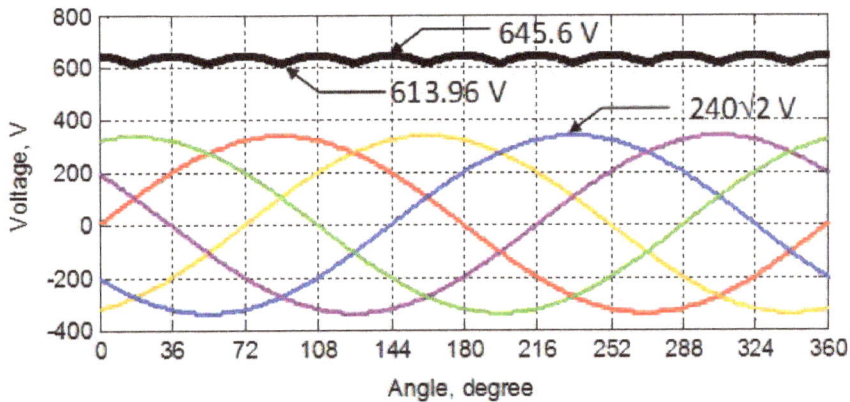

Figure 8. *Phase voltages and load voltage.*

a. The average voltage is calculated using Eq. (3):

$$V_{mean} = 1.87\,V_{ph\_max} = 1.87^*240^*\sqrt{2} = 634.7 \ V$$

b. The value of lowest harmonic of DC-side voltage is 2.1% of average DC value and occurs at 10 times the supply voltage frequency, i.e. 500 Hz:

Lowest order harmonic = $0.021^*634.7 = 13.32 \ V$

c. The maximum voltage in DC side is the peak value of line voltage for nonadjacent phases:

$$V_{max} = 1.902V_{ph\_max} = 1.902^*240\sqrt{2} = 645.5 \ V$$

The minimum voltage in DC side is the value of line voltage for nonadjacent phases at angle 18°:

$$V_{max} = 1.902V_{ph\_max}\cos 18 = 613.96 \ V$$

The percentage of ripple voltage can be calculated using Eq. (4):

$$\%Voltage \ ripple = \frac{Maximum \ value - Minimum \ value}{V_{mean}}{}^*100 = \frac{645.6 - 613.96}{634.7}{}^*100 = 4.98\%$$

d. The rms of supply current is calculated using Eq. (9):

$$I_s = 0.6324 \ I_{dc} = 15.81 \ A$$

e. The root mean square value of supply current fundamental is calculated using Eq. (8):

$$I_{s1} = 0.529 \ I_{dc} = 0.529^*25 = 13.225 \ A$$

f. The third harmonic of supply current represents 53.9% of the fundamental = 7.13 A:

The fifth harmonic of supply current = 0.

The seventh harmonic of supply current represents 23.1% of the fundamental = 3.05 A.

Example 3: *A highly inductive DC load requires 100 A at 475 V. Give design details for the diode requirement using uncontrolled five-phase full-wave rectifier.*

Solution: The required load voltage is 475 V; hence the supply voltage can be calculated using Eq. (3):

$$V_{mean} = 1.87 \ V_{ph\_max} = 475 = 1.87 \ V_{ph\_max}$$

$$\therefore V_{ph\_max} = 254 \ V$$

The required rms value of phase voltage is 179.6 ffi 180 V. The line voltage of nonadjacent phases is

$$V_L = 1.902V_{ph\_max} = 1.902^*179.6^*\sqrt{2} = 483 \ V$$

Hence, the peak reverse voltage is 483 V. Hence, each diode should withstand maximum voltage of 483 V and 100 A.

Fully controlled five-phase rectifier

The five-phase fully controlled rectifier is similar to the uncontrolled rectifier, but it has 10 thyristors instead of diodes as shown in **Figure 9**. The challenge with five-phase controlled rectifier is gating signal generation where the development and implementation of gating signals algorithm use nonadjacent line voltages. Delaying the commutation of thyristors for a certain firing delay angle 'α' controls the mean load voltage.

Figure 1 illustrates how to get DC-load voltage or mean voltage from five-phase source using five-phase controlled rectifier. **Figure 10a** shows the phase voltages, positive voltage with respect to neutral point, and negative voltage with respect to neutral point. **Figure 10b** shows the gating signals for the small firing angle α. The load voltage or output voltage can be calculated by the addition of the two five- phase half-wave voltages, relative to the supply neutral point n, appearing at the positive and negative sides of the load, respectively. As shown in **Figure 10c**, the voltage is 10 pulses in nature, labeled from '1' to '10' like uncontrolled rectifier.

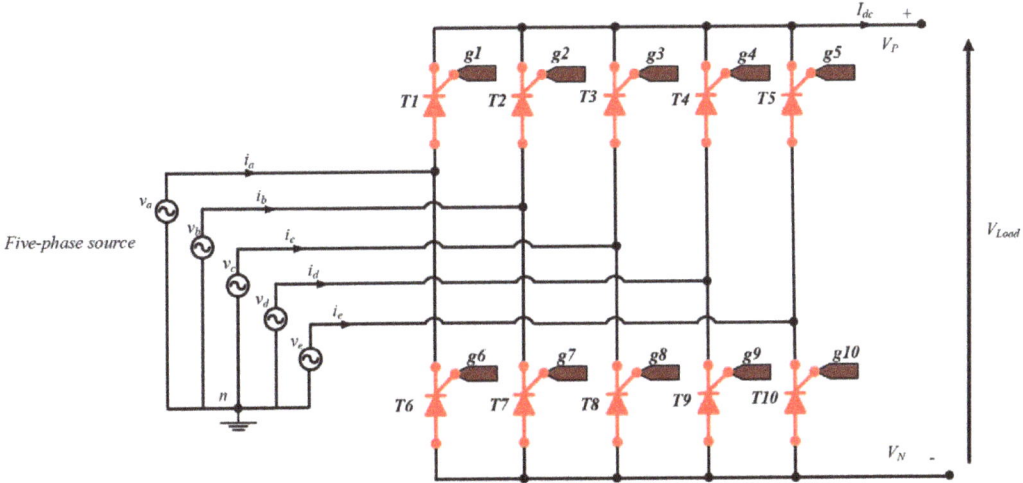

Figure 9. *Five-phase controlled line commutated rectifier.*

When v_a is the most positive phase that the thyristor, $T1$ conducts ($T1$–$T5$ are connected to the positive terminals, and $T6$–$T10$ are connected to negative terminals), and during this period, v_c is the most negative with thyristor $T8$, conducting until v_d becomes more negative, when the current in $T8$ commutates to $T9$. The next firing pulse will be to thyristor $T9$, and $T1$ is still conducting until phase b has the most positive value and is fired by $g2$, where $T9$ is still conducting, and so forth. The load voltage follows, in turn, with 10 sinusoidal voltages during one cycle, those being v_{ac}, v_{ad}, v_{bd}, v_{be}, v_{ce}, v_{ca}, v_{da}, v_{db}, v_{eb}, and v_{ec}. For the period labeled '1' in **Figure 9c**, phase a is the most positive and conducting, when $g1$ is applied to $T1$, while at the same time, phase c is conducted when $g8$ is applied to $T8$. Consequently, the load voltage ($v_{ac} = v_a - v_c$) is plotted with respect to the reference of the intersection point between phase a and phase c. Phase c is commutated with phase d for the period labeled '2', and thus, the load voltage is ($v_{ad} = v_a - v_d$) and plotted with respect to the reference of the intersection point between phase a and phase d, where the load voltage, in turn, follows the phase voltage. **Figure 10d** and **e** shows currents in $T1$ and $T6$, respectively. **Figure 10f** shows the supply current of phase 'a', which is the addition of *T1* and *T6* currents. The supply current is a quasi-square wave in nature. The supply current, the rms value of supply current, and rms value of fundamental supply current are identical to the uncontrolled rectifier case as given by Eqs. (7)–(10). **Figure 10g** shows the voltage across *T1* which can be explained using **Figure 11**.

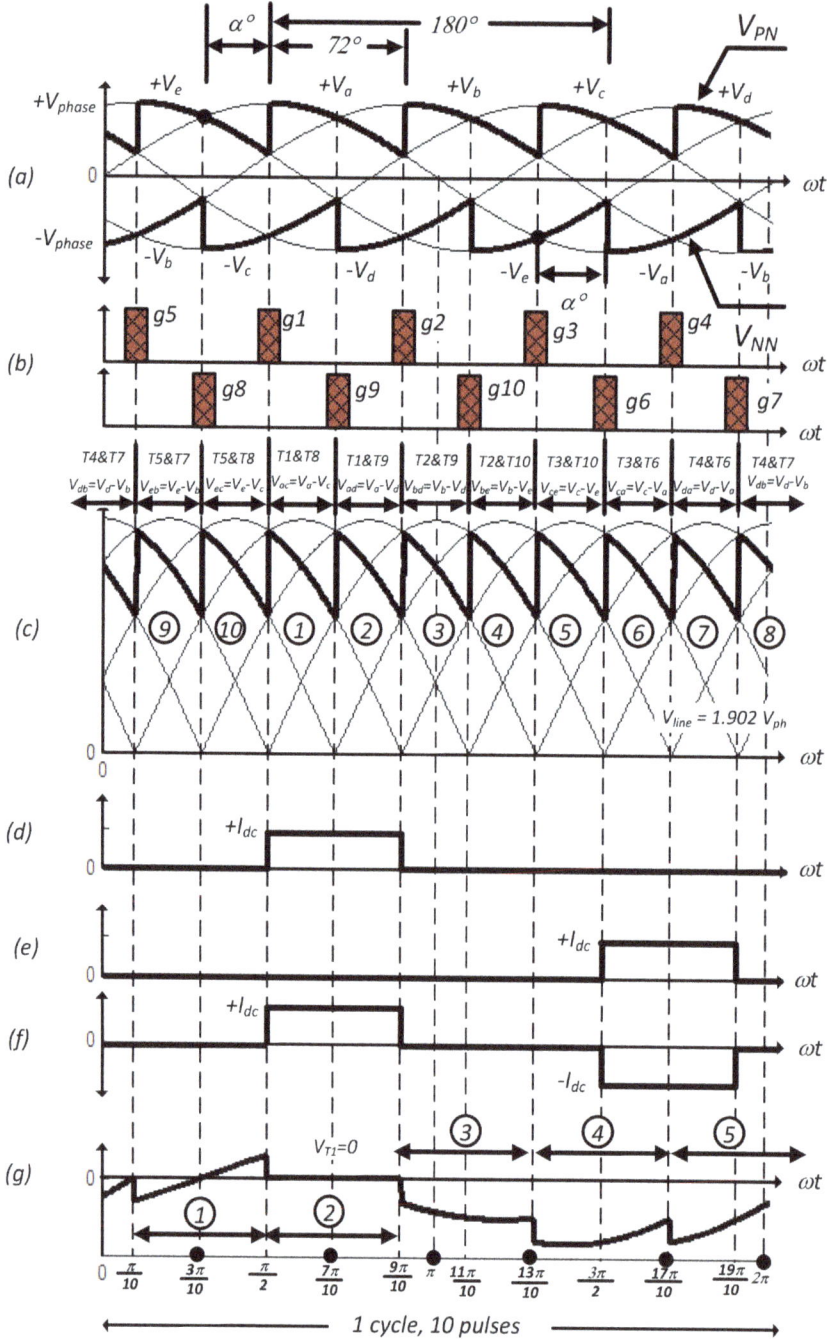

Figure 10. *Five-phase rectifier waveforms: (a) supply voltage, positive voltage with respect to neutral VPN, and negative voltage with respect to neutral VNN, (b) gating signals for firing angle α°, (c) output voltage waveform, (d) thyristor 1 current, (e) thyristor 6 current, (f) supply current for phase 'a', and (g) thyristor 1 voltage.*

The period is labeled by '2' (**Figure 10g**) and shows that the voltage across the thyristor is near zero, owing to decrease in the thyristor voltage, while the period labeled by '3' (**Figure 10g**) shows that the voltage across $T1$ can be calculated as v_{ba}, which has a maximum value of $1.1756V_{ph\_max}$, and the waveform reference point is the intersection between phases a and b. This period is explained using **Figure 11a**. The period label in 4 (**Figure 10**) is explained with the aid of **Figure 11b**. The switch has a constraint design that it should withstand during the off periods' peak reverse voltage (PRV) of $1.902V_{ph\_max}$ (line voltage for nonadjacent phases).

The mean load voltage can be evaluated with the aid of **Figures 10c** and **12** where it shows one pulse of the output load voltage:

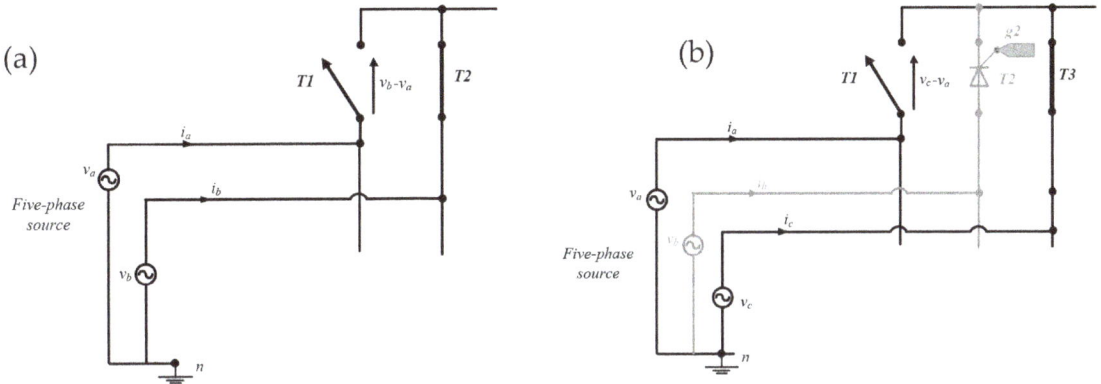

Figure 11. *Circuits calculates voltage across switch T1: (a) equivalent circuit for the period labeled 3 in* **Figure 10g** *and (b) equivalent circuit for the period labeled 4 in* **Figure 10g**.

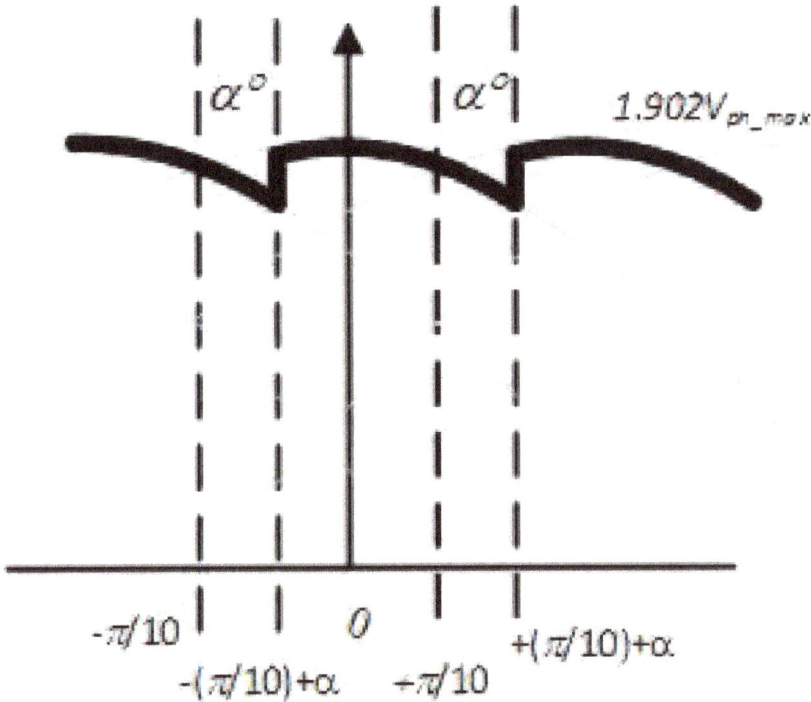

Figure 12. *One pulse of the load voltage for uncontrolled five-phase rectifier.*

$$V_{mean} = \frac{1}{2\pi/10} \int\limits_{\frac{-\pi}{10}+\alpha}^{\frac{\pi}{10}+\alpha} 1.902\, V_{ph\_max}\, \cos{(\omega t)}\, d\omega t \tag{11}$$

$$V_{mean} = 1.87 V_{ph\_max}\, \cos{(\alpha)} \tag{12}$$

The angle between the fundamental input current and fundamental input phase voltage, ϕ, is called the displacement angle. The displacement factor is defined as [3, 5, 16].

$$\text{DPF} = \cos\,\phi_1 = \cos{(\alpha)} \tag{13}$$

The supply power factor is defined as the ratio of the supply power delivered P to apparent supply power S [16]:

$$pf = \frac{P}{S} \tag{14}$$

In the case of non-sinusoidal current and voltage, the active input power to the converter delivered by the five-phase source (e.g. output of wind turbine generator) is

$$P_{in} = \frac{5}{2\pi} \int\limits_{0}^{2\pi} v_{s1}(t)\, i_{s1}(t)\, d\omega t = 5\, V_{s1}\, I_{s1}\, \cos\,\phi_1 \tag{15}$$

where V_{s1} and I_{s1} are the rms values of the fundamental components of the input phase voltage and current, respectively. The apparent power is given by

$$S = 5\, Vs\text{-}rms\; Is\text{-}rms \tag{16}$$

where $V_{s\text{-rms}}$ and $I_{s\text{-rms}}$ are the rms values of the input phase voltage and current, respectively. The displacement factor (DPF) of the fundamental current was obtained in Eq. (13). By substituting (13), (15), and (16) into (14), the power factor can be expressed as

$$pf = \frac{V_{s1} I_{s1}\, \cos\,\phi_1}{V_{s-rms} I_{s-rms}} \tag{17}$$

if the source is pure sinusoidal supply $V_{s1} = V_{s\text{-rms}}$.

The output power (load power) can be calculated using

$$P_{out} = \frac{1}{T} \int\limits_{0}^{T} v(t)\, i(t)\, dt = V_{mean} I_{mean} \tag{18}$$

The efficiency is

$$\eta = \frac{P_{out}}{P_{in}} \tag{19}$$

The total harmonic distortion, *THD*, is given by [3, 5, 15]

$$THD = \frac{\sqrt{I_s^2 - I_{s1}^2}}{I_{s1}} \tag{20}$$

For a five-phase rectifier, the *THD* is 65.5%.

Example 4: *A five-phase fully controlled rectifier is fed from 220 V (phase value), 50 Hz AC source. The DC load is highly inductive and requires 100 A at 475 V. Give design details for the thyristor requirement and the required firing angle. What is maximum DC load voltage that the converter can deliver to the load?*

Solution: The required load voltage is 475 V and the supply voltage is 220 V. The firing angle can be calculated using Eq. (12):

$$V_{mean} = 1.87\,V_{ph\_max}\cos\alpha = 475 = 1.87{}^*200{}^*\sqrt{2}\cos\alpha$$

$$\therefore \cos\alpha = 0.8998$$

$$\therefore \alpha = 26.1^\circ$$

The peak line voltage of nonadjacent phases is

$$V_L = 1.902\,V_{ph\_max} = 1.902{}^*220{}^*\sqrt{2} = 591.8\ V$$

Hence, the peak reverse voltage is $\cong$ 592 V. Hence, each thyristor should with- stand maximum voltage of 592 V and 100 A.

The maximum possible DC load voltage is when α is adjusted to zero:

$$V_{mean} = 1.87\,V_{ph\_max} = 1.87{}^*220{}^*\sqrt{2} = 581.8\ V$$

Example 5: *A five-phase fully controlled rectifier is fed from 230 V (phase value), 15 Hz AC source. The load is 45 Ω resistance and 1 H inductance (highly inductive load). Calculate for firing angle 36°, the converter power factor. Simulate your converter.*

Solution: The mean load voltage is

$$V_{mean} = 1.87\,V_{ph\_max}\cos\alpha = 1.87{}^*230{}^*\sqrt{2}\cos 36 = 492\ V$$

The load current is

$$I_{mean} = \frac{V_{mean}}{R} = \frac{492}{25} = 19.68\ \Omega$$

From Eq. (8), the rms value for fundamental supply current is 10.41 A. The angle of fundamental current is the firing angle.

From Eq. (9), the rms value of supply current is 12.44 A. The power factor is calculated using Eq. (17).

$$pf = \frac{V_{s1}I_{s1}\cos \phi_1}{V_{s-rms}I_{s-rms}} = \frac{230{}^*10.41{}^*\cos{(36)}}{230{}^*12.44} = 0.677$$

The load voltage and supply current are shown in **Figure 13**. It should be noted that as the firing angle is 36°, the load voltage shown in **Figure 13a** starts for phase 'a' at 90° which equals 54°, the intersection point between two adjacent phases, and 36° which is the firing angle. The supply current for phase 'a' is shown in **Figure 13b**. Switch 1 accommodates 72° and starts at 90°. Switch 6 is shifted 180° for the same phase.

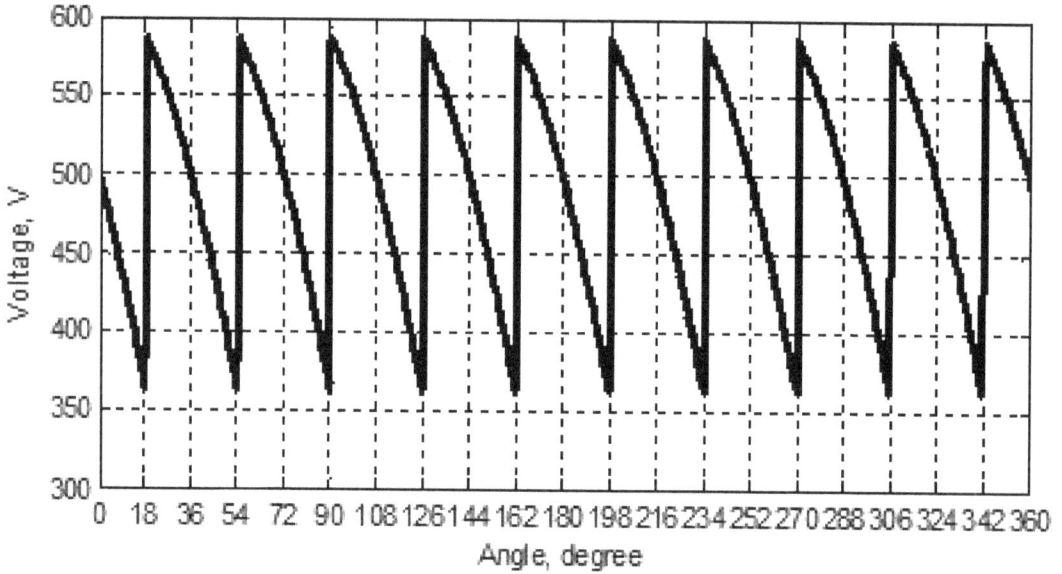

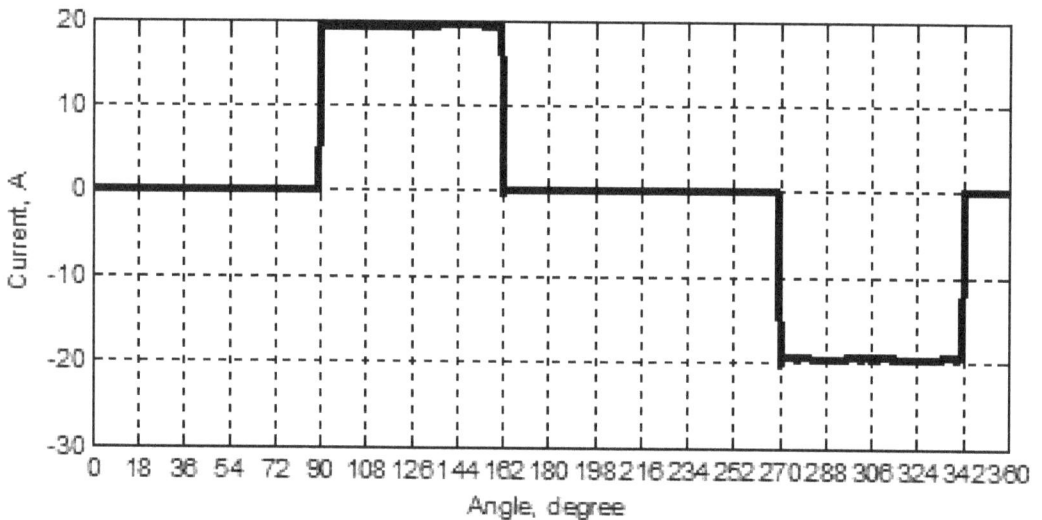

Figure 13. *Load voltage and supply current waveforms: (a) load voltage and (b) supply current.*

Gating pulse generation

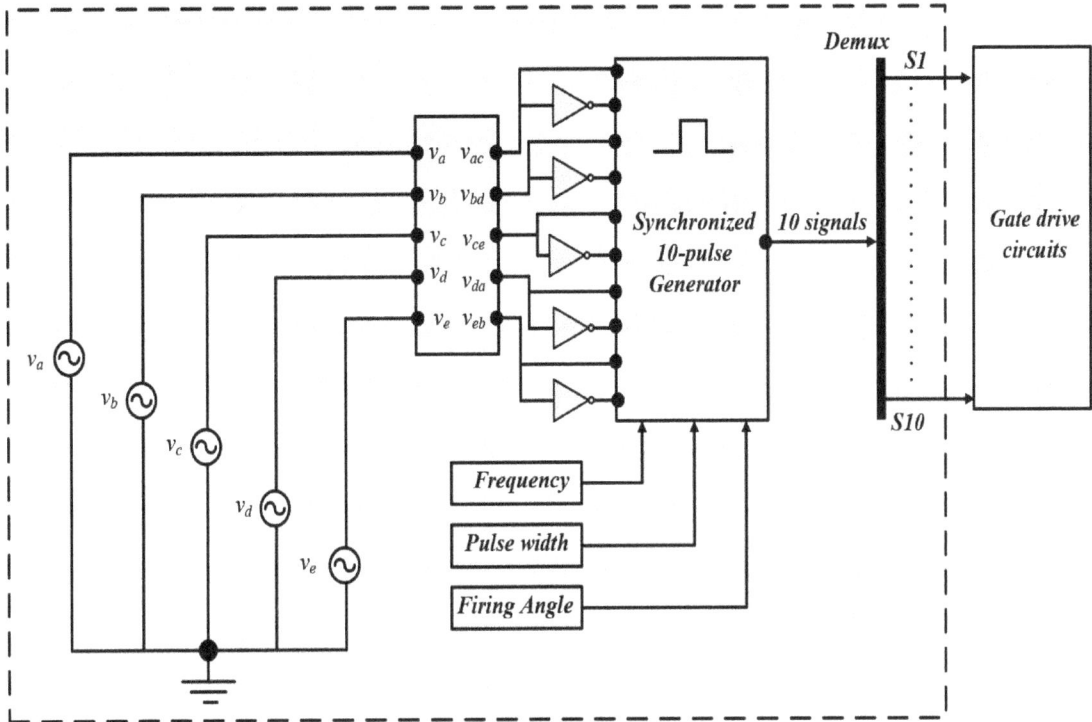

Figure 14. *Gating signal generation block.*

A pulse generator synchronized on the source voltage provides the trigger pulses for the 10 thyristors that construct the 10-pulse rectifier. The pulses are generated by α degrees after increasing zero crossings of the commutation voltages which is indicated in **Figures 10a** and **12**.

The input to the gating signal generator is 10 signals represent five-line voltages for nonadjacent phases and its inverted signals, and multiplexed, then determines the zero-crossing point and counts the delay angle α and hence applies the pulse for the specified switch through gate drive circuits as shown in **Figure 14**. A double-pulse technique is used to ensure that the specified gate signal fires its own thyristor [3, 5].

A sample of gating signal generation is shown in **Figure 15** where an example for gating signal generator output with delay angle 36° with double-pulse technique is indicated. The phase voltages of five-phase permanent magnet generator are measured as shown in part a where the rotational speed tends to have a frequency around 18 Hz and rms value of 56 V.

The delay angle is calculated from the zero-crossing point determined by the pulse generator as indicated in **Figure 15b** where gate signals 1, 2, and 6 confirm that the phase shift between gates 1 and 2 is 72° and between gates 1 and 6 is 180°. The delay angle is calculated after 54° from the start of the phase voltage [3, 5]. If the delay angle is increased to 36°, the results show that the relation between g1, g2, and g6 will be the same but shifted to 90° from the starting of phase voltage as shown in **Figure 15c**.

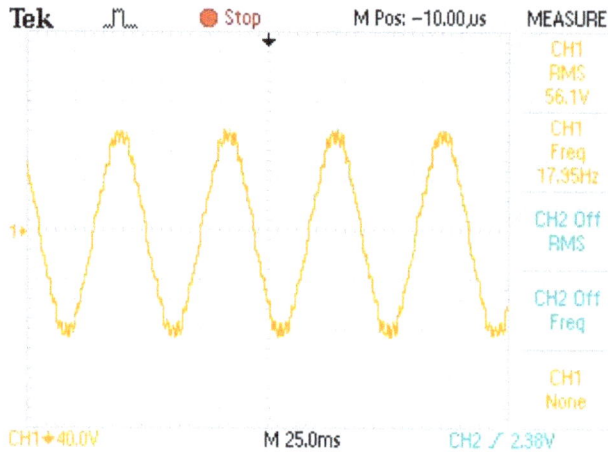

(a)

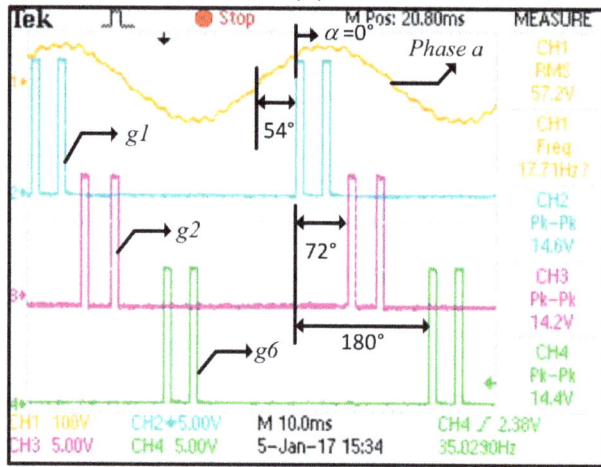

(b)

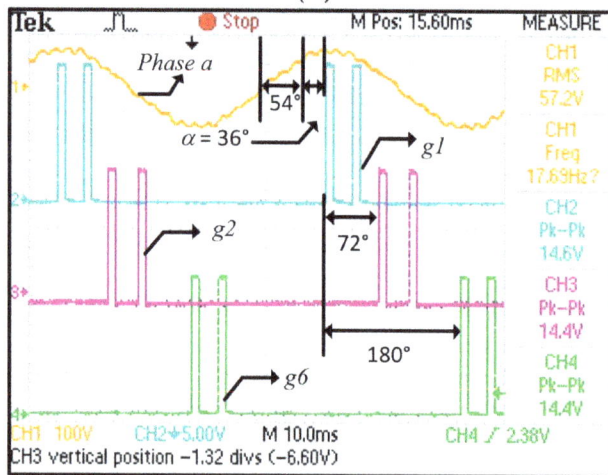

(c)

Figure 15. *Gating pulse generation: (a) Phase voltage for PM generator; (b) g1, g2, and g6 for firing angle zero; and (c) g1, g2, and g6 for firing angle 36°.*

Example 6: *For the converter given in 'Example 5', plot the gating signals*

Solution: **Figure 16** shows the gating signals for firing angle 36°.

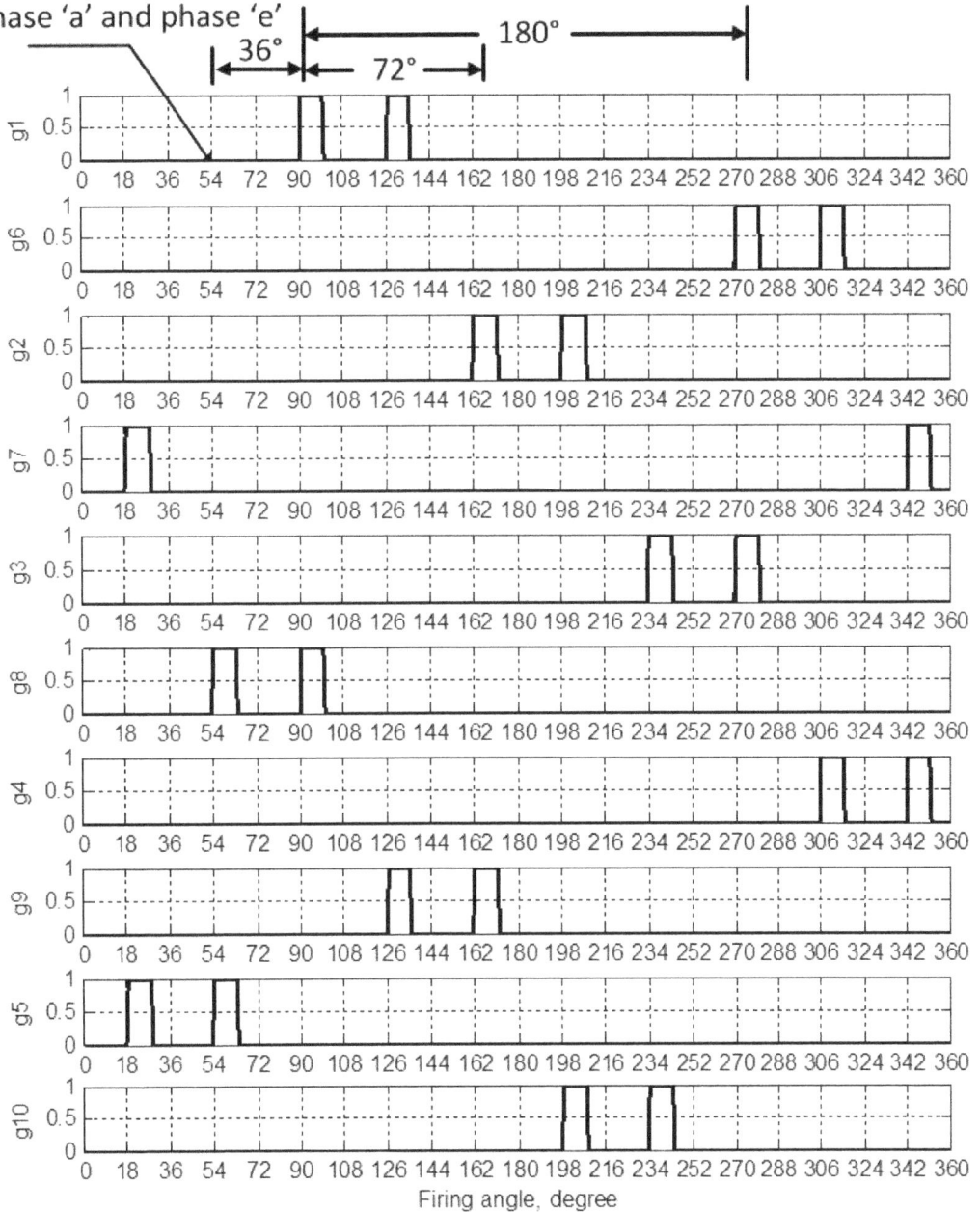

Figure 16. *Gating pulses for firing angle 36° (g1–g10).*

Conclusion

This chapter introduces the five-phase, 10-pulse, line commutated rectifier. The relation between phase voltage and line voltage for five-phase source is introduced where line voltage equals

$1.1756V_{ph\_max}$ and leads the phase voltage by 54° for adjacent phases, while it equals $1.902V_{ph\_max}$ and leads the phase voltage by 18° for nonadjacent phases. The converter performance and analysis for both AC side and DC side are detailed for both uncontrolled and controlled rectifiers. The average voltage of the DC side is 1.87 V_{ph_max} with ten pulses per one cycle, while it is 1.654 V_{ph_max} with six pulses for three-phase counterpart. Moreover, the DC-side voltage ripple is around 5%, while it is around 14% for full-wave three-phase converter. The switch (diode or thyristor) carries only 20% of the load current during one cycle, while it carries 33.3% in three-phase counterpart, and no-phase shifting transformer is required like 12-pulse converter. Also, the implementation of gating signal generation is introduced. The gating signal generation adopts double-pulse techniques and ensures the real-time synchronization between the five-phase source and gating signals. The phase shift pulse of adjacent phases is 72°, while the phase shift between pulses applied for switches on the same phase adopts 180°.

Author details

Mahmoud I. Masoud

Sultan Qaboos University, Muscat, Sultanate of Oman

*Address all correspondence to: mahmoudmasoud77@hotmail.com

References

[1] Renewables 2017 Global Status Report, REN21, Renewable Energy Policy Network. Available from: http://www.ren21.net/wp-content/uploads/ 2017/06/17-8399_GSR_2017_Full_ Report_0621_Opt.pdf [Accessed: August 8, 2018]

[2] Levi E. Multiphase electric machines for variable-speed applications. IEEE Transactions on Industrial Electronics. 2008;55(5):1893-1909. DOI: 10.1109/TIE.2008.918488

[3] Masoud MI, Saleem A, Al-Abri R. Real-time gating signal generation and performance analysis for fully controlled five-phase, 10-pulse, line- commutated rectifier. IET Transaction on Power Electronics. 2018;11(4): 744-754. DOI: 10.1049/iet-pel.2017.0587

[4] Abdel-Khalik AS, Ahmed S, Massoud AM, El-serouji A. An improved performance direct-drive Permanent Magnet wind generator using a novel single-layer winding layout. IEEE Transactions on Magnetics. 2013;49(9): 5124-5134. DOI: 10.1109/TMAG.2013.2257823

[5] Masoud MI. Fully controlled 5-phase, 10-pulse, line commutated rectifier. Alexandria Engineering Journal, AEJ. 2015;54(4):1091-1104. DOI: 10.1016/j.aej.2015.07.004. Elsevier

[6] Abdel-Khalik AS, Masoud MI, Williams BW. Eleven phase induction machine: Steady-state analysis and performance evaluation with harmonic injection. The Institution of Engineering Technology, IET, Electric Power Applications, EPA. 2010;4(8):670-685. DOI: 10.1049/iet-epa.2009.0204

[7] Wu B. High Power Converters and AC Drives. Piscataway, NJ: IEEE Press; 2006. DOI: 10.13140/RG.2.1.1093.2967

[8] Lander CW. Power Electronics. 3rd ed. England: McGraw-Hill; 1993. ISBN: 0-07-707714-8

[9] Iqbal A, Payami S, Saleh M, Anad A, Behra RK. Analysis and control of a five- phase AC/DC/AC converter with active rectifier. In: The 22nd Australasian Universities Power Engineering Conference (AUPEC); Bali, Indonesia; 2012. pp. 1-6

[10] Blaabjerg F, Ma K. Future on power electronics for wind turbine systems. IEEE Journal of Emerging and Selected Topics in Power Electronics. 2013;1(3): 139-152. DOI: 10.1109/ JESTPE.2013.2275978

[11] Masoud MI, Al-Abri RS. Five-phase current source PWM rectifier. In: The 7th IEEE GCC International Conference and Exhibition; GCC, Doha, Qatar; 2013. pp. 148-152

[12] Rodriguez JR, Dixon LW, Espinoza JR, Pontt J, Lezana P. PWM regenerative rectifiers: State of the art. IEEE Transactions on Industrial Electronics. 2005;52(1):5-22. DOI: 10.1109/ TIE.2004.841149

[13] Abdelsalam AK, Masoud MI, Finney SJ, Williams BW. Medium- voltage pulse width modulated current source rectifiers using different semiconductors: Loss and size comparison. International Journal of Engineering technology. IET, Transaction on Power Electronics. 2010; 3(2):243-258. DOI: 10.1049/iet-pel.2008.0252

[14] Naseri F, Haidar S. A comparison study of high power IGBT-based and Thyristor-based AC to DC converters in medium power dc arc furnace plants. In: The 9th International Conference in Compatibility and Power Electronics; Lisbon, Portugal; 2015. pp. 14-19

[15] Rashid M. Power Electronics Circuits, Devices, and Applications. 3rd ed. USA: Prentice Hall; 2004. ISBN-10: 0131011405

[16] Tzeng YS. Harmonic analysis of parallel-connected 12-pulse uncontrolled rectifier without an inter-phase transformer. IEE Proceedings of Electric Power Applications. 1998; 145(3):253-260. DOI: 10.1049/ip-epa: 19981850

Permissions

All chapters in this book were first published in EPC, by InTech Open; hereby published with permission under the Creative Commons Attribution License or equivalent. Every chapter published in this book has been scrutinized by our experts. Their significance has been extensively debated. The topics covered herein carry significant findings which will fuel the growth of the discipline. They may even be implemented as practical applications or may be referred to as a beginning point for another development.

The contributors of this book come from diverse backgrounds, making this book a truly international effort. This book will bring forth new frontiers with its revolutionizing research information and detailed analysis of the nascent developments around the world.

We would like to thank all the contributing authors for lending their expertise to make the book truly unique. They have played a crucial role in the development of this book. Without their invaluable contributions this book wouldn't have been possible. They have made vital efforts to compile up to date information on the varied aspects of this subject to make this book a valuable addition to the collection of many professionals and students.

This book was conceptualized with the vision of imparting up-to-date information and advanced data in this field. To ensure the same, a matchless editorial board was set up. Every individual on the board went through rigorous rounds of assessment to prove their worth. After which they invested a large part of their time researching and compiling the most relevant data for our readers.

The editorial board has been involved in producing this book since its inception. They have spent rigorous hours researching and exploring the diverse topics which have resulted in the successful publishing of this book. They have passed on their knowledge of decades through this book. To expedite this challenging task, the publisher supported the team at every step. A small team of assistant editors was also appointed to further simplify the editing procedure and attain best results for the readers.

Apart from the editorial board, the designing team has also invested a significant amount of their time in understanding the subject and creating the most relevant covers. They scrutinized every image to scout for the most suitable representation of the subject and create an appropriate cover for the book.

The publishing team has been an ardent support to the editorial, designing and production team. Their endless efforts to recruit the best for this project, has resulted in the accomplishment of this book. They are a veteran in the field of academics and their pool of knowledge is as vast as their experience in printing. Their expertise and guidance has proved useful at every step. Their uncompromising quality standards have made this book an exceptional effort. Their encouragement from time to time has been an inspiration for everyone.

The publisher and the editorial board hope that this book will prove to be a valuable piece of knowledge for researchers, students, practitioners and scholars across the globe.

List of Contributors

Brij Mohan Mundotiya
National Institute of Technology (NIT), Hamirpur, Himachal Pradesh, India

Wahdat Ullah and Krishna Kumar
Malaviya National Institute of Technology (MNIT), Jaipur, Rajasthan, India

Mohamed Ahmed Ebrahim
Faculty of Engineering at Shoubra, Benha University, Cairo, Egypt

R.G. Mohamed
Eastern Company, Giza, Egypt

Marian Găiceanu
Dunarea de Jos University of Galati, Galati, Romania
Integrated Energy Conversion Systems and Advanced Control of Complex Processes Research Center, Galati, Romania

Carlos A. Reusser
Department of Electronics, Universidad Tecnica Federico Santa Maria, Valparaiso, Chile

Ivan Ramljak and Drago Bago
J. P. Elektroprivreda HZ HB d.d Mostar, Bosnia and Herzegovina

Dana Alghool, Noora Al-Khalfan, Stabrag Attiya and Farayi Musharavati
Department of Mechanical and Industrial Engineering, College of Engineering, Qatar University, Doha, Qatar

Mohammed Salem
School of Electrical and Electronic Engineering, Universiti Sains Malaysia, Nibong Tebal, Malaysia

Khalid Yahya
Kocaeli University, İzmit, Kocaeli, Turkey

Thaiyal Naayagi Ramasamy
School of Electrical and Electronic Engineering, Newcastle University International Singapore, Singapore

Boxue Du, Jin Li and Hucheng Liang
Key Laboratory of Smart Grid of Education Ministry, School of Electrical and Information Engineering, Tianjin University, Tianjin, China

Kiran Singh
Department of Electrical Engineering, Indian Institute of Technology, Roorkee, Uttarakhand, India

Mahmoud I. Masoud
Sultan Qaboos University, Muscat, Sultanate of Oman

Index

www.ingramcontent.com/pod-product-compliance
Lightning Source LLC
Chambersburg PA
CBHW062001190326
41458CB00009B/2931